Applied Mathematical Sciences

EDITORS

Fritz John
Courant Institute of
Mathematical Sciences
New York University
New York, N.Y. 10012

Lawrence Sirovich
Division of
Applied Mathematics
Brown University
Providence, R.I. 02912

Joseph P. LaSalle
Division of
Applied Mathematics
Brown University
Providence, R.I. 02912

Gerald B. Whitham
Applied Mathematics
Firestone Laboratory
California Institute of Technology
Pasadena, CA. 91109

EDITORIAL STATEMENT

The mathematization of all sciences, the fading of traditional scientific boundaries, the impact of computer technology, the growing importance of mathematical-computer modelling and the necessity of scientific planning all create the need both in education and research for books that are introductory to and abreast of these developments.

The purpose of this series is to provide such books, suitable for the user of mathematics, the mathematician interested in applications, and the student scientist. In particular, this series will provide an outlet for material less formally presented and more anticipatory of needs than finished texts or monographs, yet of immediate interest because of the novelty of its treatment of an application or of mathematics being applied or lying close to applications.

The aim of the series is, through rapid publication in an attractive but inexpensive format, to make material of current interest widely accessible. This implies the absence of excessive generality and abstraction, and unrealistic idealization, but with quality of exposition as a goal.

Many of the books will originate out of and will stimulate the development of new undergraduate and graduate courses in the applications of mathematics. Some of the books will present introductions to new areas of research, new applications and act as signposts for new directions in the mathematical sciences. This series will often serve as an intermediate stage of the publication of material which, through exposure here, will be further developed and refined and appear later in one of Springer-Verlag's other mathematical series.

MANUSCRIPTS

The Editors welcome all inquiries regarding the submission of manuscripts for the series. Final preparation of all manuscripts will take place in the editorial offices of the series in the Division of Applied Mathematics, Brown University, Providence, Rhode Island.

SPRINGER-VERLAG NEW YORK INC., 175 Fifth Avenue, New York, N.Y. 10010

Printed in U.S.A.

Applied Mathematical Sciences | Volume 16

S. Lefschetz

Applications of Algebraic Topology

Graphs and Networks
The Picard-Lefschetz Theory and Feynman Integrals

With 52 Illustrations

Springer-Verlag New York · Heidelberg · Berlin
1975

S. Lefschetz

Formerly of Princeton University

AMS Classifications: 55-01, 55A15, 81A15

Library of Congress Cataloging in Publication Data

Lefschetz, Solomon, 1884-1972.
Applications of algebraic topology.

(Applied mathematical sciences; v. 16)
Bibliography: p.
Includes index.
1. Algebraic topology. 2. Graph theory. 3. Electric networks. 4. Feynman integrals. I. Title. II. Series.
QA1.A647 vol. 16 [QA611] 510'.8 [514'.2] 75-6924

ISBN 978-0-387-90137-4 ISBN 978-1-4684-9367-2 (eBook)
DOI 10.1007/978-1-4684-9367-2

Solomon Lefschetz (1884-1972) was one of the great mathematicians of his generation. This volume published posthumously and completed shortly before his death is in his own unique and vigorous style. Were he alive there are many people whom he would thank. Among them are Sandra Spinacci for the careful typing of his manuscript, Mauricio Peixoto for his constant encouragement, and John Mallet-Paret for his careful reading of the manuscript.

January 1975 J. P. LaSalle

TABLE OF CONTENTS

PART I

APPLICATION OF CLASSICAL TOPOLOGY TO GRAPHS AND NETWORKS

		Page
INTRODUCTION		4
CHAPTER I.	A RÉSUMÉ OF LINEAR ALGEBRA	5
	1. Matrices	5
	2. Vector and Vector Spaces	7
	3. Column Vectors and Row Vectors	10
	4. Application to Linear Equations	11
CHAPTER II.	DUALITY IN VECTOR SPACES	13
	1. General Remarks on Duality	13
	2. Questions of Nomenclature	14
	3. Linear Functions on Vector Spaces. Multiplication	15
	4. Linear Transformations. Duality	16
	5. Vector Space Sequence of Walter Mayer	18
CHAPTER III.	TOPOLOGICAL PRELIMINARIES	22
	1. First Intuitive Notions of Topology	22
	2. Affine and Euclidean Spaces	25
	3. Continuity, Mapping, Homeomorphism	26
	4. General Sets and Their Combinations	27
	5. Some Important Subsets of a Space	28
	6. Connectedness	29
	7. Theorem of Jordan-Schoenflies	30
CHAPTER IV.	GRAPHS. GEOMETRIC STRUCTURE	34
	1. Structure of Graphs	34
	2. Subdivision. Characteristic Betti Number	37
CHAPTER V.	GRAPH ALGEBRA	43
	1. Preliminaries	43
	2. Dimensional Calculations	46
	3. Space Duality. Co-theory	48
CHAPTER VI.	ELECTRICAL NETWORKS	51
	1. Kirchoff's Laws	51
	2. Different Types of Elements in the Branches	53
	3. A Structural Property	54
	4. Differential Equations of an Electrical Network	56
CHAPTER VII.	COMPLEXES	61
	1. Complexes	61
	2. Subdivision	64
	3. Complex Algebra	66
	4. Subdivision Invariance	68
CHAPTER VIII.	SURFACES	71
	1. Definition of Surfaces	71
	2. Orientable and Nonorientable Surfaces	72
	3. Cuts	76
	4. A Property of the Sphere	79

		Page
	5. Reduction of Orientable Surfaces to a Normal Form	83
	6. Reduction of Nonorientable Surfaces to a Normal Form	84
	7. Duality in Surfaces	86
CHAPTER IX.	PLANAR GRAPHS	89
	1. Preliminaries	89
	2. Statement and Solution of the Spherical Graph Problem	90
	3. Generalization	95
	4. Direct Characterization of Planar Graphs by Kuratowski	96
	5. Reciprocal Networks	103
	6. Duality of Electrical Networks	104

PART II

THE PICARD-LEFSCHETZ THEORY AND FEYNMAN INTEGRALS

INTRODUCTION		113
CHAPTER I.	TOPOLOGICAL AND ALGEBRAIC CONSIDERATIONS	119
	1. Complex Analytic and Projective Spaces	119
	2. Application to Complex Projective n-space $\mathcal{P}^n$	119
	3. Algebraic Varieties	121
	4. A Résumé of Standard Notions of Algebraic Topology	124
	5. Homotopy. Simplicial Mappings	128
	6. Singular Theory	129
	7. The Poincaré Group of Paths	130
	8. Intersection Properties for Orientable M^{2n} Complex	131
	9. Real Manifolds	133
CHAPTER II.	THE PICARD-LEFSCHETZ THEORY	135
	1. Genesis of the Problem	135
	2. Method	136
	3. Construction of the Lacets of Surface Φ_z	138
	4. Cycles of Φ_z. Variations of Integrals Taken On Φ_z	140
	5. An Alternate Proof of the Picard-Lefschetz Theorem	140
	6. The Λ_1-manifold M. Its Cycles and Their Relation to Variations	146
CHAPTER III.	EXTENSION TO HIGHER VARIETIES	149
	1. Preliminary Remarks	149
	2. First Application	150
	3. Extension to Multiple Integrals	151

			Page
	4.	The 2-Cycles of an Algebraic Surface	152
CHAPTER IV.		FEYNMAN INTEGRALS	154
	1.	On Graphs	154
	2.	Algebraic Properties	156
	3.	Feynman Graphs	160
	4.	Feynman Integrals	162
	5.	Singularities	163
	6.	Polar Loci	164
	7.	More General Singularities	168
	8.	On the Loop-Complex	170
	9.	Some Complements	170
	10.	Examples	171
	11.	Calculation of an Integral	174
	12.	A Final Observation	175
CHAPTER V.		FEYNMAN INTEGRALS. B.	177
	1.	Introduction	177
	2.	General Theory	177
	3.	Relative Theory	178
	4.	Application to Graphs	178
	5.	On Certain Transformations	180
BIBLIOGRAPHY			181
SUBJECT INDEX PART I			183
SUBJECT INDEX PART II			187

PART I

APPLICATION OF CLASSICAL TOPOLOGY TO GRAPHS AND NETWORKS

PREFACE

This monograph is based, in part, upon lectures given in the Princeton School of Engineering and Applied Science. It presupposes mainly an elementary knowledge of linear algebra and of topology. In topology the limit is dimension two mainly in the latter chapters and questions of topological invariance are carefully avoided.

From the technical viewpoint graphs is our only requirement. However, later, questions notably related to Kuratowski's classical theorem have demanded an easily provided treatment of 2-complexes and surfaces.

January 1972 Solomon Lefschetz

INTRODUCTION

The study of electrical networks rests upon preliminary theory of graphs. In the literature this theory has always been dealt with by special ad hoc methods. My purpose here is to show that actually this theory is nothing else than the first chapter of classical algebraic topology and may be very advantageously treated as such by the well known methods of that science.

Part I of this volume covers the following ground: The first two chapters present, mainly in outline, the needed basic elements of linear algebra. In this part duality is dealt with somewhat more extensively. In Chapter III the merest elements of general topology are discussed. Graph theory proper is covered in Chapters IV and V, first structurally and then as algebra. Chapter VI discusses the applications to networks. In Chapters VII and VIII the elements of the theory of 2-dimensional complexes and surfaces are presented. They are applied in Chapter IX, the last of Part I, to the important question of planar graphs, Kuratowski related theorem, and dual networks.

It is to be noted that in the electrical part, linearity has nowhere been assumed. In general as regards networks, I have been considerably inspired by the splendid paper of Brayton and Moser: A theory of nonlinear networks, Quaterly of Applied Mathematics, Vol. 29 pp. 1-33, 81-104, 1964.

The exposition of the material is new in many parts; moreover in certain parts the material is completely new. This is notably the case in Chapter IX.

CHAPTER I

A RÉSUMÉ OF LINEAR ALGEBRA

Two elements dominate linear algebra: matrices and vectors. One may identify vectors with certain matrices but not vice versa. Thus matrices are the dominant feature. We shall, therefore, first deal with matrices and then with vectors.

As appropriate for a résumé, proofs will rarely be given and for them the reader is referred to any standard text on the subject.

1. Matrices

A matrix is a rectangular array of elements

$$\begin{bmatrix} a_{11} & a_{12} & \cdots & a_{1n} \\ a_{21} & a_{22} & \cdots & \cdot \\ \cdots & \cdots & \cdots & \cdots \\ a_{m1} & a_{m2} & \cdots & a_{mn} \end{bmatrix}.$$

Such an array, known as $m \times n$ matrix is usually abridged as $[a_{jk}]$ or even written a. The standard matrix operations are:

Addition: The sum of two $m \times n$ matrices, a as above and $b = [b_{jk}]$ is the matrix

$$a + b = [a_{jk} + b_{jk}];$$

Product: With a as before and b an $n \times p$ matrix one defines

$$a\, b = \left[\sum_s a_{js}\, b_{sk}\right].$$

The implication is that in both addition and multiplication the operations indicated have a meaning. This is usually clear from the context but one must not be entirely careless about it.

The transpose a' of the $m \times n$ matrix a is the $n \times m$ matrix obtained by permuting the rows and columns of a. Note that if ab has a meaning $(ab)' = b'a'$.

The derivative of a matrix $a(t) = [a_{jk}(t)]$ of elements differentiable functions of t is

$$\dot{a}(t) = [\dot{a}_{jk}(t)].$$

Square matrices. These are the $n \times n$ matrices. The number n is the order of the matrix.

A square numerical, $n \times n$ matrix has a determinant written $|a_{jk}|$ or $|a|$. The matrix is singular if $|a| = 0$, nonsingular otherwise.

The square matrix with diagonal $a_1, \ldots, a_n$ and zeros outside is frequently written $\mathrm{diag}(a_1, \ldots, a_n)$. The unit matrix of order n, written E_n or E (when n is obvious) is $\mathrm{diag}(1,1,\ldots,1)$ (n terms).

A nonsingular matrix a has an inverse a^{-1} characterized by $aa^{-1} = a^{-1}a = E$. If $|a| \neq 0$, $|b| \neq 0$, then $(ab)^{-1} = b^{-1}a^{-1}$.

Recall this important property: inversion and transposition commute. That is $(a^{-1})' = (a')^{-1}$.

Evidently, sums and products of $n \times n$ matrices are $n \times n$ matrices.

Rank of a matrix. The rank ρ of an $m \times n$ numerical matrix a is the largest order of a nonzero determinant formed from the elements of a.

(1.1) Theorem. Let a be an $m \times n$ matrix and b, c nonsingular square matrices of respective order m, n. Then $\mathrm{rank}\ a = \mathrm{rank}\ b\,a\,c$.

It is convenient to note that if $a = [a_{jk}]$ is $m \times n$ and $b = \text{diag}(b_1,\ldots,b_m)$, $c = \text{diag}(c_1,\ldots,c_n)$.

Then

$$ba = [b_j\, a_{jk}], \quad ac = [a_{jk}\, c_k].$$

2. Vectors and Vector Spaces

Vectors are inextricably mixed with a collection of numbers, the scalars, called a field. A field is simply any set of elements obeying the ordinary rules of rational operations, for example all real or all complex numbers. However an interesting field is made up of just two elements 0 and 1 under these rules:

$$0.0 = 0.1 = 1.0 = 1 + 1 = 0; \; 1.1 = 1.$$

In that field, called the field mod 2, $x = -x$, $\frac{1}{x} = x$, $(x \neq 0)$ hence subtraction and division may be forgotten. This is the ideal field in geometric questions in which direction does not occur.

Take now a fixed field F and n elements $A_1,\ldots,A_n$ which obey no special relation (pure symbols). Form all the expressions

$$A = \alpha_1\, A_1 + \cdots + \alpha_n\, A_n$$

with coefficients in F, the obvious rule for addition and the conventions $A = 0$ if every $\alpha_h = 0$, likewise

$$\alpha A = (\alpha\alpha_1)\, A_1 + \cdots + (\alpha\alpha_n)\, A_n$$

for every α in F. The collection of all expressions A is a vector space V, the elements A are the vectors.

The vectors $B_1, B_2,\ldots,B_p$ are linearly dependent if there exists a relation

$$\beta_1\, B_1 + \cdots + \beta_r\, B_r = 0, \; \beta_h \text{ in } F$$

with the β_h not all zero (non-trivial relation). If no such relation exists the B_h are linearly independent (the term "linearly" is often omitted in such statements). The maximum number of linearly _independent_ vectors is the _dimension_ of V

$$\dim V = n. \tag{2.1}$$

Bases. A base for the space V is a set $B_1, \ldots, B_s$ of independent vectors such that every vector C satisfies a relation

$$C = \beta_1 B_1 + \cdots + \beta_s B_s, \ \beta_h \ \text{in} \ F.$$

(2.2) _A base consists exactly of_ $n(=\dim V)$ _elements_.

(2.3) _Any_ n _independent elements form a base_. _Hence_ $A_1, \ldots, A_n$ _is a base_.

Isomorphism. Two vector spaces V,W over the same field F are isomorphic, written $V \sim W$, if there is a one-one correspondence between their elements preserving the relations of dependence between them. That is if $B_1, \ldots, B_s$ are elements of V and C_h corresponds to B_h then the relations

$$\sum \beta_h B_h = 0, \quad \sum \beta_h C_h = 0$$

imply one another.

(2.4) _N.a.s.c. to have_ $V \sim W$ _is that they have the same dimension_.

(2.5) _If_ $V \sim W$ _one may select for them respective bases_ $\{B_h\}, \{C_h\}$ _such that the isomorphism between them associates_ $\sum \beta_h B_h$ _with_ $\sum \beta_h C_h$.

Change of base. Let $\{B_h\}, \{C_h\}$ be two bases for the same vector space V. We have the relations

$$C_h = \sum \gamma_{hj} B_j, \quad B_h = \sum \beta_{hj} C_j$$

with the β,γ in the field F. As a consequence there follow

$$B_h = \sum_s \beta_{hs} \gamma_{sk} B_k, \quad h = 1,2,\ldots,n. \tag{2.6}$$

However, since the B_h are independent these relations must be identically true, that is

$$\sum_s \beta_{hs} \gamma_{sk} = \begin{cases} 1 & \text{if } h = k \\ 0 & \text{otherwise} \end{cases}.$$

This means that the product

$$[\beta_{hj}] \cdot [\gamma_{hj}] = E \tag{2.7}$$

and implies for the determinants

$$|\beta_{hj}| \cdot |\gamma_{hj}| = 1.$$

Consequently, the matrices $[\beta_{hj}]$ and $[\gamma_{hj}]$ are non-singular. Conversely any relation

$$C_h = \sum \gamma_{hj} B_j, \quad |\gamma_{hj}| \neq 0 \tag{2.8}$$

is a change of base from $\{B_h\}$ to $\{C_h\}$ for the space V.

Remark. The important properties of the space V are those which are invariant with respect to changes of base. For the present we only have the dimension, but other properties will appear in the application to graphs.

Direct sum. Let V_1, V_2 be two vector subspaces of V (vector spaces over the same field whose vectors are all in V). We say that V is their direct sum and write

$$V = V_1 \oplus V_2$$

whenever the following two conditions hold:

(a) V_1 and V_2 only have zero in common;

(b) if B is any vector of V then $B = B_1 + B_2$, where B_h is in V_h. Then also

$$\dim V = \dim V_1 + \dim V_2. \tag{2.9}$$

(2.10) If V_1 is a subspace of V, there is another subspace V_2 such that $V = V_1 \oplus V_2$. The subspace V_2 need not be unique, but all such subspaces are isomorphic. By identifying them in a suitable manner there results a unique space V_2 called the factor space of V by V_1 and written V/V_1.

3. Column Vectors and Row Vectors

Let the numbers $x_1, x_2, \ldots,$ be elements of the scalar field F and consider all the 1-column matrices

$$x = \begin{bmatrix} x_1 \\ x_2 \\ \vdots \\ x_n \end{bmatrix},$$

under the addition rule for $n \times 1$ matrices, the multiplication rule $\alpha x = \text{diag}(\alpha, \ldots, \alpha) \cdot x$, α in F, and the convention $x = 0$ if and only if every $x_h = 0$, then the collection of all matrices x makes up a vector space V. Its dimension is n, hence it is isomorphic with the vector space V_0 of elements

$$x_1 A_1 + \cdots + x_n A_n$$

already defined in Section 2. We may think of V_0 as a representation of the space V. In this context one refers to x as a column-vector, and to the x_h as components of x. The transpose matrix $x' = [x_1,\ldots,x_n]$ is called a <u>row-vector</u>. The space V_r of all the x' is again a representation of V. It is worth noting that if one calls A the formal $n \times 1$ matrix

$$A = \begin{bmatrix} A_1 \\ A_2 \\ \vdots \\ A_n \end{bmatrix}$$

then one may write

$$x_1 A_1 + \cdots + x_n A_n = x' A = A'x.$$

These abridged designations will be found most convenient in later chapters. Notice, also that if x and y are both column vectors with n terms all taken from the field F then

$$x'y = y'x = \sum x_h y_h. \tag{3.1}$$

This is the well known direct product of the vectors x,y.

If the space of the x_h is considered as Euclidean, with coordinates x_h then

$$x'x = ||x||^2 \quad \text{(Euclidean length square)}.$$

4. Application to Linear Equations

Let

$$a_{j1} x_1 + \cdots + a_{jn} x_n = 0 \tag{4.1}$$
$$(j = 1,2,\ldots,m)$$

be a system of linear equations with coefficients in a field F. If $a = [a_{jk}]$ and x denotes the column vector $[x_h]$ then (4.1) is the same as the equation

$$ax = 0. \tag{4.2}$$

Let r be the rank of the matrix a. Then from the well known elementary theory of equations we have:

(4.3) <u>The solution vectors of the system (4.2) make up a vector space of dimension</u> $n - r$.

Consider also the system

$$y'a = 0 \tag{4.4}$$

where y is an m-vector. Since this equation is equivalent to $a'y = 0$, and a, a' have the same rank, we have from (4.3)

(4.5) <u>The solution vectors of (4.4) make up a vector space of dimension</u> $m - r$.

Exercises

1. Derive the proofs missing in 1, 2, 3, and 4.

2. Let $P = [p_{ij}]$ be a real $m \times n$ matrix. The rank r of P is the largest order of a determinant extracted from P. Prove that if Q is $m \times m$, R is $n \times n$ both nonsingular then rank P = rank QPR.

3. Let P be real $n \times n$ and symmetric: $P' = P$. If x is any vector $x'Px = \Phi(x)$. Show that by a linear transformation of coordinates $x = Py$ one may reduce Φ in various ways to form $\Sigma\, \alpha_h y_h^2$ where the α_h are all real. Show that in such a reduction:

(a) the number of $\alpha \neq 0$ is fixed and equal to the rank of P;

(b) the number of positive α_h is likewise fixed.

This number is known as the index of inertia of P. (This last result is due to Sylvester.)

CHAPTER II

DUALITY IN VECTOR SPACES

1. General Remarks on Duality

The idea of duality occurs in many parts of mathematics. Its earliest appearance (some 125 years ago) was in projective geometry where it permitted to halve the number of theorems to be proved. It also played a most important role in analysis, for example in Banach spaces. In topology, beginning with Poincaré its role has been no less important.

The central attack on duality in modern mathematics may be described in these terms. If S is a space of any sort on which one may specify linear functions, then their space Σ is defined as the dual space to S.

It is our purpose to develop duality for vector spaces from the standpoint just described.

Observe that the relatively simple duality of projective geometry fits in perfectly with the above general description. To be precise consider the projective plane P_x with related projective coordinates x_1, x_2, x_3, under these conditions:

(a) the three coordinates are never simultaneously zero;

(b) the point (kx_1, kx_2, kx_3), $k \neq 0$, is the same as (x_1, x_2, x_3).

A linear function of the point is an expression $\phi(x) = u_1 x_1 + u_2 x_2 + u_3 x_3$. If one excludes the form $\phi \equiv 0$ and identifies ϕ and $k\phi$, $k \neq 0$, then the points (u_1, u_2, u_3) of the projective plane P_u are in one-one correspondence with the linear forms ϕ. The plane P_u is the <u>dual</u> of the plane P_x. Observe finally that the form ϕ is completely identified by the point (u_1,u_2,u_3) of the plane P_u. Thus lines of P_x correspond to points of P_u and points of P_x to the lines of P_u. Thus the duality relation $P_x \longleftrightarrow P_u$ is entirely symmetrical.

2. Questions of Nomenclature

We shall generally accept the following standard designations:

I. Transformations: A transformation f from one collection G or elements to another H is written $f\colon G \to H$.

II. Kronecker deltas: These are the symbols $\delta_{hk} = 0$ for $h \neq k$, $= 1$ for $h = k$.

III. Dimension: This term is generally dropped and one says "n-plane, space, sphere,... "for" n-dimensional plane, space, sphere,...".

IV. n-vector stands for a vector in a vector n-space.

V. Vector spaces and their bases. The general designation for these spaces is by Latin capitals A,B,C... . A vector space, say A, will usually be referred explicitly or implicitly to a definite base. Assume that $\dim A = n$ and that it has the base $\underline{e}^{(1)},\underline{e}^{(2)},\ldots,\underline{e}^{(n)}$. An element of A will be an expression $a_1\underline{e}^{(1)} +\cdots+ a_n\underline{e}^{(n)}$, where a_h are its components. As a result the space A is represented by the space of the column-vector $\underline{a} = [a_h]$. That is there is a tacit identification of the space A with the space of column vectors $[a_h]$.

The preceding situation occurs frequently in geometry. For instance in plane geometry referred to the axes x_1,x_2 one will speak of "the point (x_1,x_2)" meaning actually "the point presently represented in this coördinate system by the members x_1,x_2".

Incidentally, if one thinks of the vectors of A as "points"

the space A is also known as affine n-space.

3. Linear Functions on Vector Spaces. Multiplication

Let A be a vector space and let its base be $\{\underline{e}^{(1)},\underline{e}^{(2)},\ldots,\underline{e}^{(n)}\}$. A function $f(\underline{a})$ on A to the reals is linear whenever given any two elements $\underline{a},\underline{a}^1$ of A and any two real scalars, α,β:

$$f(\underline{a}+\underline{a}^1) = f(\underline{a}) + f(\underline{a}^1).$$

It follows that if

$$\underline{a} = a_1\underline{e}^{(1)} + \cdots + a_n\underline{e}^{(n)}$$

then

$$f(\underline{a}) = a_1a_1^* + a_2a_2^* + \cdots + a_na_n^* \tag{3.1}$$

$$a_h^* = f(\underline{e}^{(h)}),$$

so that the a_h^* are all real. Suppose that $g(\underline{a})$ is another linear function on A to the reals and let

$$g(\underline{a}) = a_1b_1^* + \cdots + a_nb_n^* .$$

If we define $\alpha f + \beta g$ by the relation

$$(\alpha f+\beta g)(\underline{a}) = \alpha f(\underline{a}) + \beta g(\underline{a})$$

for every $\underline{a}$ of A, then the linear functions f will become the elements of a vector space A^* whose base is $(\underline{e}^{*(1)},\ldots,\underline{e}^{*(n)})$, where $\underline{e}^{*(h)}$ is the particular function defined by

$$\underline{e}^{*(h)}(\underline{e}^{(k)}) = \delta_{hk}.$$

(The δ_{hk} are the Kronecker deltas: 1 if $h = k$, 0 otherwise.)

The space A^* is the dual of the space A, and since the base of A^* consists of n terms $\dim A^* = \dim A = n$.

One may write (3.1) as

$$\underline{a}^*(\underline{a}) = \sum a_h^* a_h \ . \tag{3.2}$$

It is clear from (3.2) that $A^{**} = A$: duality of spaces is a symmetrical relationship.

One may also aptly write (3.2) as a <u>multiplication</u>

$$\underline{a}^* \cdot \underline{a} = \sum a_h^* a_h$$

and we recognize that <u>this new product is commutative</u>.

The new product has been extensively utilized in the literature. In particular, in algebraic topology it has been referred to as <u>Kronecker index</u>.

4. Linear Transformations. Duality

Let B be a second vector space. A transformation $\phi: A \to B$ is linear whenever given $\underline{a}, \underline{a}'$ of A and any two real scalars α, α' we have

$$\phi(\alpha\underline{a}+\alpha'\underline{a}') = \alpha\phi(\underline{a}) + \alpha'\phi(\underline{a}').$$

Let $\{\underline{e}^{(h)}\}$ and $\{\underline{f}^{(k)}\}$ be bases for A and B. Then

$$\phi(\underline{e}^{(h)}) = \sum \eta_{hk}\underline{f}^{(k)}.$$

Hence if $\underline{a} = \sum a_h \underline{e}^{(h)}$, then

$$\phi(\underline{a}) = \sum a_h \eta_{hk} \underline{f}^{(k)}.$$

That is in the column-vector representation

$$\phi(\underline{a}) = \underline{a}'\eta. \tag{4.1}$$

If $n = \dim A$ and $m = \dim B$ the matrix η is $n \times m$.

The <u>nucleus</u> $N(\phi)$ of ϕ consists of all the vectors $\underline{a}$ sent by ϕ into zero. These vectors are characterized by

$$\sum a_h \eta_{hk} \underline{f}^{(k)} = 0. \tag{4.2}$$

Since the $\underline{f}^{(k)}$ satisfy no relation from (4.2) follows that

$$\sum_h a_h \eta_{hk} = 0,$$

that is in a vector notation

$$\underline{a}'\eta = 0. \tag{4.3}$$

<u>This is the characteristic equation of the elements of the nucleus</u> $N(\phi)$ <u>of</u> ϕ.

Let B^* be the dual of B. We propose to define a dual linear transformation $\phi^*: B^* \to A^*$.

Observe first this general property:

(4.4) <u>An identity in</u> x, y

$$f(\underline{x},\underline{y}) = \sum \xi_{hk} x_h y_k \equiv 0$$

<u>where</u> $\xi = [\xi_{hk}]$ <u>is a constant matrix, is equivalent to</u> $\xi = 0$.

It is clear that $\xi = 0$ implies $f = 0$. Conversely $f \equiv 0$ implies $f(\delta_{h1}, \delta_{h2}, \ldots, \delta_{hn}, \delta_{k1}, \delta_{k2}, \ldots, \delta_{km}) = \xi_{hk} = 0$ that is $\xi = 0$.

We shall now prove:

(4.5) <u>The relations</u> $\zeta' = \eta$ <u>and</u>

$$\phi^*(\underline{b}^*)\underline{a} = \phi(\underline{a})\underline{b}^* \tag{4.6}$$

<u>are equivalent where</u> $\phi^*(\underline{f}^{*(k)}) = \sum \zeta_{kh} \underline{e}^{*(h)}$.

In fact

$$\phi^*(\underline{b}^*)\underline{a} = \underline{b}^{*\prime}\zeta\underline{a}$$

$$\phi(\underline{a})\underline{b}^* = \underline{a}'\eta\underline{b}^* = \underline{b}^{*\prime}(\underline{a}'\eta)' = \underline{b}^{*\prime}\eta'\underline{a}.$$

Hence (4.6) is equivalent to

$$\underline{b}^{*'}(\zeta-\eta')\underline{a} \equiv 0$$

and therefore to $\zeta = \eta'$ by (4.4).

Whenever either (4.6) holds or equivalently $\zeta' = \eta$ we shall say that ϕ^* is dual to ϕ. This implies

$$\phi \text{ is dual to } \phi^* \quad \text{(symmetry of duality).} \tag{4.7}$$

An important consequence of (4.5) is:

(4.8) *A n.a.s.c. in order that* a *be in the nucleus* $N(\phi)$ *of* ϕ *is that* $\phi^*(\underline{b}^*)\underline{a} = 0$ *for all* b^*. *Similarly a n.a.s.c. to have* b^* *in the nucleus* $N(\phi^*)$ *is that* $\phi(\underline{a})\underline{b}^* = 0$ *for all* a.

It is sufficient to treat the first case. Necessity being obvious let the condition hold. Then by (4.6), $\phi(\underline{a})\underline{b}^* = 0$ for all $\underline{b}^*$. Hence all the coefficients of this linear form in the b_h^* must vanish, and this implies that $\phi(\underline{a}) = 0$: $\underline{a}$ is in $N(\phi)$.

5. Vector Space Sequence of Walter Mayer

Consider a finite sequence of vector spaces $0, A_n, A_{n-1}, \ldots, A_1, 0$ with linear transformations $\phi_p\colon A_p \to A_{p-1}$.

The assumption, borrowed from topology, is made that

$$\phi_{p-1}\phi_p = 0, \text{ all } p. \tag{5.1}$$

These are the Walter Mayer sequences.

Special notation. Before proceeding let us introduce a convenient notation. If spaces are such that $P = Q \oplus R$, and $R \cong S$ we shall write $P = Q \oplus S$. Since our main concern is to find certain dimensional relations, we merely note that we continue to have $\dim P = \dim Q + \dim S$.

The relation (5.1) has noteworthy consequences. First if Z_p is the nucleus of ϕ_p then there is a space D_p such that

$$A_p = D_p \oplus Z_p \ . \tag{5.2}$$

Then if $F_{p-1} = \phi_p A_p$ it follows from (5.1) that

$$\phi_{p-1}\phi_p A_p = \phi_{p-1}F_{p-1} = 0.$$

That is F_{p-1} is a subspace of Z_{p-1}, or equally F_p is a subspace of Z_p. Hence there is a H_p such that

$$Z_p = F_p \oplus H_p \tag{5.3}$$

so that

$$A_p = D_p \oplus F_p \oplus H_p \ . \tag{5.4}$$

The relation (5.2) also implies that the nucleus of ϕ_{p+1} as a transformation $D_{p+1} \to F_p$ is zero. Hence ϕ_{p+1} is an isomorphism $D_{p+1} \to F_p$. Thus finally

$$A_p = F_{p-1} \oplus F_p \oplus H_p \ . \tag{5.5}$$

Let α_p, r_p, R_p be the dimensions of A_p, F_p, H_p. Then (5.5) yields

$$R_p = \alpha_p - r_{p-1} - r_p \tag{5.6}$$

and consequently

$$\sum (-1)^p R_p = \sum (-1)^p \alpha_p \ . \tag{5.7}$$

These expressions are well known from Topology. Borrowing from that science one may refer to the R_p, and to (5.7) as <u>Betti numbers</u> and <u>characteristic</u> of the Mayer system $\{A_p, \phi_p\}$.

In order to calculate R_p we require the numbers r_p. We prove

r_p <u>is the rank of the matrix</u> ζ_p <u>of</u> ϕ_p. (5.8)

The components of the element $\underline{z}^{(p)}$ of Z_p are defined by

$$\underline{z}^{(p)'}\zeta_p' = 0.$$

Hence $\dim Z_p = \alpha_p - r_p'$, where r_p' is the rank of ζ_p. Since

$$A_p = D_p \oplus Z_p$$

we have

$$r_p = \dim D_p = \alpha_p - \dim Z_p = \alpha_p - (\alpha_p - r_p') = r_p' .$$

Thus r_p is equal to the rank of the matrix ζ_p.

<u>Dual relations</u>. One defines the dual spaces A^*, B^* and the rest in the obvious way.

The sequence of the duals is

$$0 \to A_1^* \to \cdots \to A_p^* \xrightarrow{\phi_{p+1}^*} A_{p+1}^* \to \cdots \to A_n^* \to 0$$

behave exactly like the initial sequence. Denoting by $Z_p^*,\ldots,$ the obvious dual groups we have again

$$A_p^* = D_p^* \oplus Z_p^* ,$$

$$A_p^* \doteq F_p^* \oplus F_{p+1}^* \oplus H_p^* ,$$

$$F_p^* = \phi_p^*(D_{p-1}^*).$$

Since the matrices are merely transposed the ranks are preserved. Since likewise $\dim A_p^* = \alpha_p$ we have again $\dim H_p^* = R_p$.

Thus $\dim H_p^* = \dim H_p$.

We also have from (4.8) with $A = A_p$ and $B = A_{p-1}$:

$$\underline{a}_{p-1}^{*'} \cdot \phi_p(\underline{a}_p) = \phi_p^*(\underline{a}_{p-1}^*)' \cdot \underline{a}_p. \tag{5.9}$$

(5.10) A n.a.s.c. to have $\underline{a}_p$ in Z_p is that $\phi_p^*(\underline{a}_{p-1}^*)'\underline{a}_p = 0$ for every $\underline{a}_{p-1}^*$. Similarly, a n.a.s.c. to have z_{p-1}^* in Z_{p-1}^* is that $\underline{z}_{p-1}^{*\prime}\phi_p(\underline{a}_p) = 0$ for every $\underline{a}_p$.

CHAPTER III

TOPOLOGICAL PRELIMINARIES

There are many approaches to topology. One of the most accessible is by means of the notion of distance. Our purpose in the present chapter is to sketch this approach and a few of the general concepts derivable from it. It may be said that this material amply covers our future needs.

1. First Intuitive Notions of Topology

Take some very thin wire and with it make up the five objects sketched in Figure 1. While they seem to be

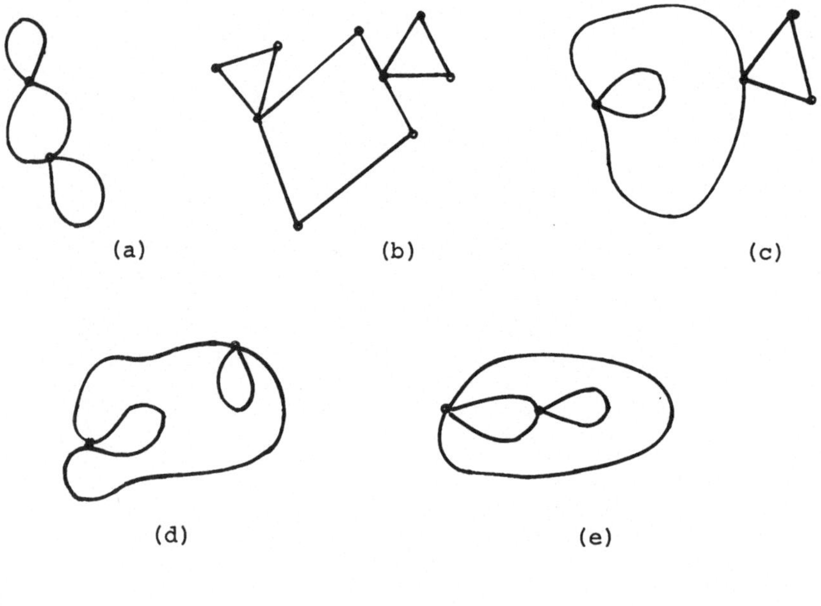

(a) (b) (c)

(d) (e)

Figure 1.

very different it is obvious that one may continuously modify any one of them into any other point for point. One may say the same thing about the three shaded areas sketched in Figure 2. Similar comparisons

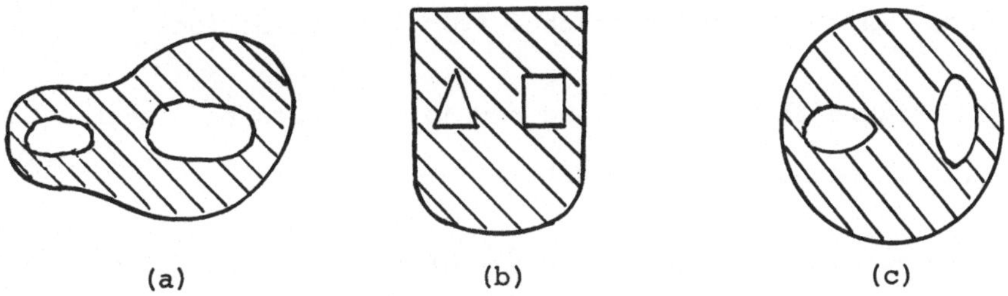

Figure 2.

hold for a sphere, an egg shell, a tetrahedron (Figure 3). The sort of transition envisaged in each of these three groups of figures is

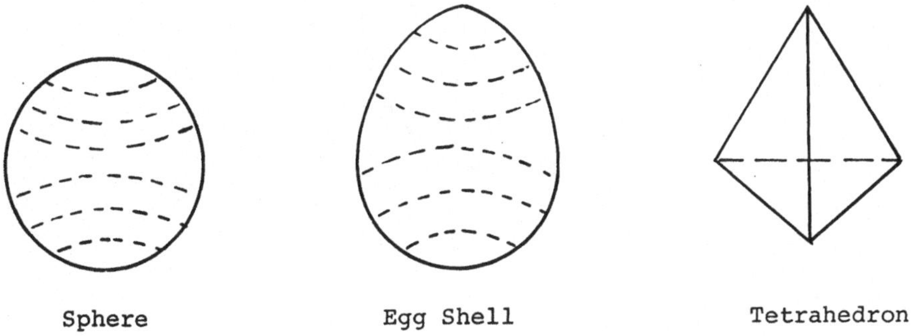

Figure 3.

said to be <u>topological</u> or a <u>homeomorphism</u> (defined with precision in Section 3). <u>Topology</u> is concerned with the properties common to all the figures, for example in Groups 1, 2, 3. Are there any common properties in each of our three groups of figures? Certainly, and, some at least, are very simple.

<u>Group 1</u>. Figure 1: In each there are two and exactly two points (the nodes) which may be approached from four directions.

<u>Group 2</u>. Figure 2: In each, one may join by arcs the inner boundaries to the outer boundaries.

<u>Group 3</u>. Take any circuit on any one of the figures of the group and cur the surface along the circuit -- this is known as a <u>cut</u>; as a consequence each of the figures breaks up into two disjoint pieces.

On the contrary take a torus T (Figure 4). One may draw

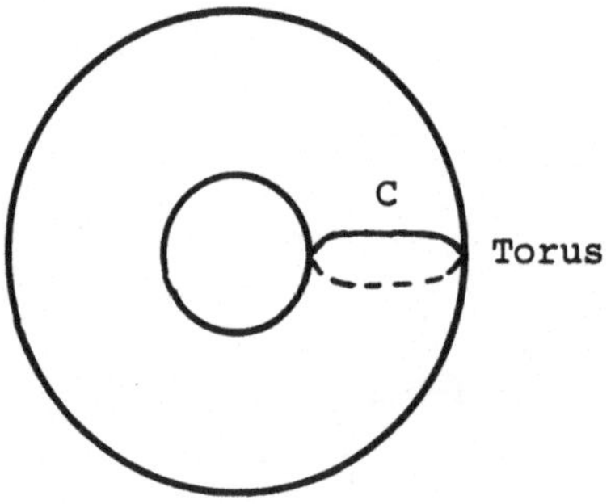

Figure 4.

on T a cut C which does not disconnect T. This shows that the torus and sphere cannot be modified in our manner into one another.

A curious and very famous figure may be obtained from a rectangle ABCD (Figure 5a). Upon bridging the sides AB and CD

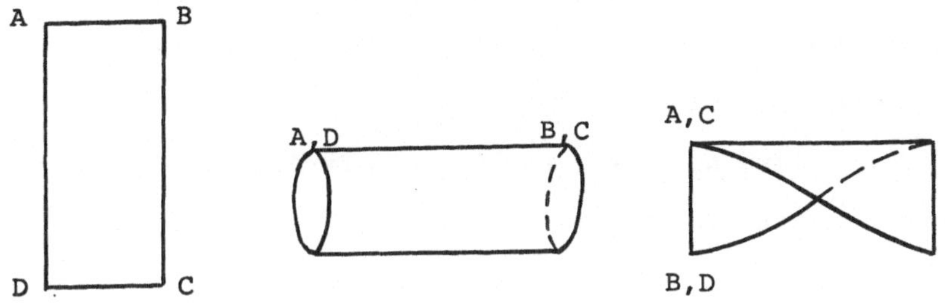

(a) (b) (c)

Möbius Strip
Figure 5.

into coincidence so that D coincides with A and C with B one just obtains the tube of Figure 5b. However, if the coincidence is of A with C and D with B there results the <u>Möbius strip</u> of Figure 5c. This figure was introduced by Möbius to generate a <u>non-orientable</u> surface: If a point P is followed by a transverse arrow as it describes a circuit on the strip then as P returns to its initial position the arrow may have been reversed. That figures b and c are not transformable into one another by our earlier process is easily verified by observing that Figure 5b has two disjoint boundary curves and Figure 5c has only one. The two figures are not homeomorphic.

2. Affine and Euclidean Spaces

The intuitive description of topology which has been given is not adequate for a solid mathematical theory. A firmer foundation will now be provided. Since we shall ultimately only be concerned with Euclidean figures it is necessary to say a few words about affine and Euclidean spaces.

An affine n-space $\mathscr{A}^n$ is just a collection of points each determined by n real numbers $x_1,\ldots,x_n$ the coordinates of the point. It is convenient to identify the point with these coordinates with the column vector x which has the components x_h, and we shall merely say "the point x".

The affine n-space $\mathscr{A}^n$ becomes Euclidean n-space $\mathscr{E}^n$ by merely assigning to its point pairs x,y the distance

$$d(x,y) = (\Sigma(x_h-y_h)^2)^{1/2}.$$

Once this is done the only allowed coordinate transformations are those preserving this distance.

One may easily verify these basic properties:

I. $d(x,y) \geq 0$, and $= 0$ if and only if $x = y$;

II. (Triangle Law): $d(x,y) + d(y,z) \geq d(x,z)$;

III. (Symmetry): $d(x,y) = d(y,x)$.

Our spaces. They will consist solely of vector spaces and of Euclidean figures. Such a figure F, immersed in $\mathscr{E}^n$ will then possess a distance $d(x,y)$ equal to the distance of x,y in $\mathscr{E}^n$. In particular, this distance will possess properties I, II, III.

3. Continuity, Mapping, Homeomorphism

With distances at our disposition we are able to give accurate definitions for some of our fundamental concepts.

Let F and G be two spaces. Let the letter d cover equally their distance functions. Let ϕ be a transformation or function on $F \to G$. Let x of F be sent by ϕ to $\phi x = y$ of G. If every y of G is a ϕx we say: ϕ is onto, or ϕ covers G;

otherwise that ϕ is into. We also describe ϕ as <u>continuous at the point</u> x whenever given any positive number ζ there is another η such that if $y' = \phi x'$ then the requirement $d(y,y') < \zeta$ is met by imposing $d(x,x') < \eta$. Or in words, but less precisely: to have y' near enough to y it is sufficient to take x' near enough to x. A transformation ϕ which is continuous at all points of the space F is known as a <u>mapping</u>. Examples of mappings are folding of a square F around a diagonal ($\phi: F \to F$), or projection of a circular region F into a line G of its plane ($\phi: F \to G$).

Let again ϕ be merely a transformation $F \to G$. If every point y of G comes from a unique x through ϕ then there is defined a unique transformation $\phi: G \to F$ written ϕ^{-1} and called <u>inverse</u> of ϕ. We say then that ϕ is 1 - 1. A <u>topological</u> transformation or <u>homeomorphism</u> $\phi: F \to G$ is a transformation which is 1 - 1 and bicontinuous: both ϕ and ϕ^{-1} are continuous.

The <u>topology</u> of the space F is the study of all the properties of F, which persist under a homeomorphism. These properties are said to be <u>topological</u>.

4. General Sets and Their Combinations

Generally speaking a <u>set</u> is just any collection of objects, called points for convenience, so that one speaks of a <u>point-set</u>. We have also used the term <u>figure</u> for Euclidean sets. There are a few standard combinations of sets and associated symbols that may be utilized later. In describing them the letters A,B will refer to general sets of elements:

$A \supset B$ or $B \subset A$: B is a subset of A;

$A + B$: <u>union</u> of A and B or set of elements in A or in B;

$A \cdot B$: or AB: <u>intersection</u> of A and B or set of elements in both A and B;

A - B: if $B \subset A$ it is the set of all points in A but not in B, called, in general, complement of B in A;

A - B = A - AB: complement of AB in A.

5. Some Important Subsets of a Space

Let our space R be a definite subset of some Euclidean space. All the subsets to be mentioned are to be subsets of R.

Spheroid $\mathcal{S}(x,r)$ of center x and radius r: set of all points nearer than r to x;

Open set U: any union of spheroids. One agrees also that vacuum is (formally) an open set.

Neighborhood N(x) of a point x: any open set containing x.

Closed set F: complement R - U of an open set.

If $A \subset R$ then $A\mathcal{S}(x,r)$, is a spheroid in A of the point x of A, and UA,FA are an open and a closed set of A.

Examples. Let R be a Euclidean plane $\mathcal{E}^2$. Then the circular or polygonal regions are open sets of R, while a line, an ellipse are closed sets of R.

(5.1) Cells and spheres. These are two figures of frequent occurrence later and contributing important topological types.

A zero-cell is just a point. For $n > 0$ we have: an open n-cell is the homeomorph of the Euclidean set

$$x'x = x_1^2 + x_2^2 + \cdots + x_n^2 < 1.$$

Replacing $<$ by $\leq$ yields the closed n-cell.

The (n-1)-sphere is the homeomorph of the set of $\mathcal{E}^n$ represented by $x'x = 1$. The zero-sphere consists of just two points.

A one-cell is called arc, a closed one-cell is called closed arc.

An interval is an open one-cell on a line. A segment is the corresponding closed one-cell.

Standard designations are: (a,b) for the interval $a < t < b$ and $[a,b]$ for the segment $a \leq t \leq b$ ($a < b$ throughout).

Let ϕ be a topological mapping $[a,b] \to F$ with $x = \phi a$, $y = \phi b$ and $\lambda = \phi[a,b]$. We say then that x,y are joined by the arc λ in F.

6. Connectedness

This is an important topological property of constant occurrence later. The definition given presently is a restricted version of a more general definition. It is, however, ample for our purpose.

Let F be a figure. Two points x,y of F are connected if they may be joined by an arc in F. The set $C(x)$ of all points y of F which are connected with x is the component of x. Let y of $C(x)$ be connected with z so that z is in $C(y)$. Denote by λ,μ arcs joining x to y and y to z. Let λ followed from x

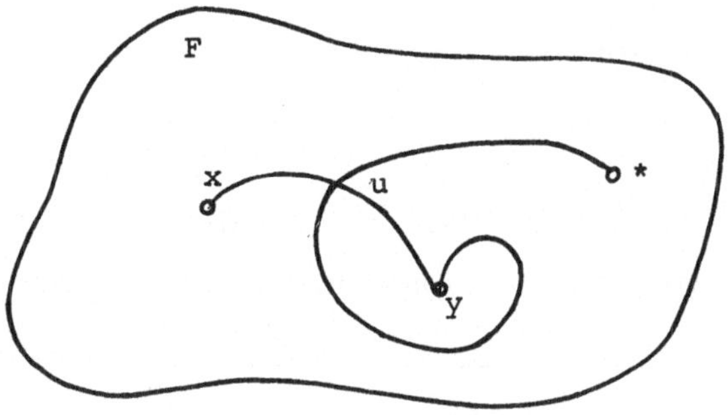

Figure 6.

to y first meet μ at the point u. Let λ' be the part of λ from x to u and μ' the part of μ from u to z (μ followed after u). Then $\lambda' + \mu'$ is an arc connecting x to z. Hence $C(y) \subset C(x)$ and vice versa, and so $C(x) = C(y)$. That is the component of every point of $C(x)$ is $C(x)$ itself. For this reason we refer to $C(x)$ as a component of the figure F itself. F is said to be connected if it has just one component (namely itself): any two points are connected by an arc.

Examples: n-cells and n-spheres $n > 0$, are connected. One agrees also that a point is connected. The type of connectedness here considered is sometimes designated as arcwise connectedness.

Under a homeomorphism connected sets go into connected sets. Hence the number $\rho(F)$ of components of F is a topological invariant of F.

7. Theorem of Jordan-Schoenflies

This is a classical theorem of topology required repeatedly later. As the proof is definitely arduous no attempt will be made to give it here.

A Jordan curve is merely a 1-sphere, that is the homeomorph of a circle.

(7.1) Theorem. A Jordan curve J situated on a 2-sphere S divides S into two component regions U,V with the common boundary J, and $U + J$, $V + J$ are both closed 2-cells. Equivalently: the same property holds for a Jordan curve in a plane Π save that only one of the regions is a 2-cell and the other is infinite.

The assertion that $S - J$ consists of two regions with the common boundary J was first stated, but defectively proved by Camille Jordan about a century ago -- hence the name "Jordan curve".

This part of the theorem is often referred to as "Jordan curve theorem". Its correct proof was first given by Oswald Veblen (Chicago thesis 1904). The Schoenflies' part refers to the assertion that $U + J$ and $V + J$ are 2-cells. A proof of the Jordan curve theorem is given in the author's Introduction to Topology, p. 65.

It is evident that all these results are topological. We shall accept them without proof.

The following, also given without proof is a reasonable exercise.

(7.2) *Let* U *be one of the 2-cells of* J. *Then an arc* λ *in* U *with endpoints distinct and in* J *divides* U *into two 2-cells* U_1, U_2 *whose common boundary in* U *is* λ.

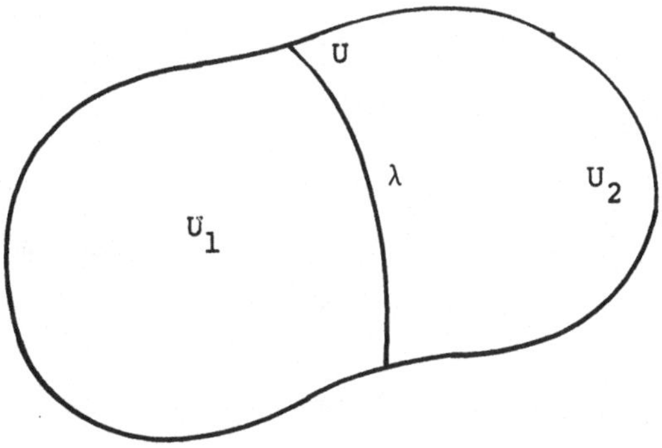

Figure 7.

Exercises

In the following exercises we have collected a number of interesting topological properties. Although not required later the reader may find them worth perusing.

1. Let $\phi: F \to G$ be a transformation. Show that the continuity of ϕ is equivalent to the following property: the points of any open set V of G comes from an open set U of F.

Consequences: (a) a homeomorphism $\phi: F \longleftrightarrow G$ is characterized by mere interchange of open sets: (b) open sets and closed sets are topological invariants.

2. Prove that the union of any number and the intersection of a finite number of open sets are open sets. Show also that the union of a finite number and the intersection of any number of closed sets are closed sets.

3. Let R be a space and $A \subset R$. A limit point of A is a point x of R such that every $\mathscr{S}(x,r)$ contains points of $A - x$.

Show that closed sets F of R are characterized by this property: limit-points of any subset of F are points of F.

4. The closure $\bar{A}$ of a set A is the intersection of all closed sets $\supset A$ (least closed set over A). Show that $\bar{A}$ is the set of all points at distance zero from A. Hence a closed set is characterized by the property $F = \bar{F}$.

5. Given a set of real numbers N let $\sup N$ denote the least upper bound of all numbers $\geq N$ and $\inf N = -\sup(-N)$.

The distance $d(x,A)$ from a point x to a set A is defined as $\inf d(x,y)$ for all points y of A; the diameter $d(A)$ of A is $\sup d(x,y)$ for all x,y in A.

Show that $\bar{A}$ is the set of all points at zero distance from A.

6. Compactness. An open covering of R is any collection of open sets containing R. The space R is compact if every open covering has a finite subcovering. Show the equivalence of this property: every non-empty subset of R has a limit point.

7. A compact set is bounded (containing in some spheroid) and closed.

8. Let $f(x)$ be a continuous real function of the point x of a compact space R. Then $f(x)$ attains both $\sup f(x)$ and $\inf f(x)$ on R.

9. The compact subsets of E^n are exactly those which are closed and bounded.

10. A union of connected sets with common point is connected.

11. A sequence of connected sets $A_1, A_2, A_3, \ldots$, such that $A_h A_{h+1}$ is never empty has a connected union.

12. Let $F_1, F_2, \ldots$ be bounded closed connected sets in E^m such that $F_1 \supset F_2 \supset F_3 \ldots$. Then the intersection $F_1 F_2 F_3 \ldots$ is closed non-empty connected (or a point) and bounded.

13. A more general definition of connectedness than the one given in Section 6 is this: The set A is connected if it is impossible to have $A = B + C$ where B, C are disjoint, non-empty and open in A (hence, also closed in A). Prove that under this definition

(a) a closed arc is connected.

(b) Show that the properties of 10, 11, 12 also hold under this more general definition.

14. Prove the Jordan-Schoenflies theorem for a plane polygon.

CHAPTER IV

GRAPHS. GEOMETRIC STRUCTURE

The properties of a finite graph (only type considered) may be divided into two distinct groups: geometric, really topological properties, and algebraic properties. In the present chapter, we present the geometry of a graph and in the next its algebra.

1. Structure of Graphs

By definition a finite graph G consists of a finite collection of points: its nodes $n_1, n_2, \ldots, n_{\alpha_0}$ and disjoint arcs $b_1, b_2, \ldots, b_{\alpha_1}$ the branches of G. We assume that each branch has two distinct end points which are nodes; and also that every node is an endpoint of some branch: and that no two branches have the same endpoints.

Order of a point. A point x belongs to a certain number of closed branches of G. Let each of these be cut by removal of one or

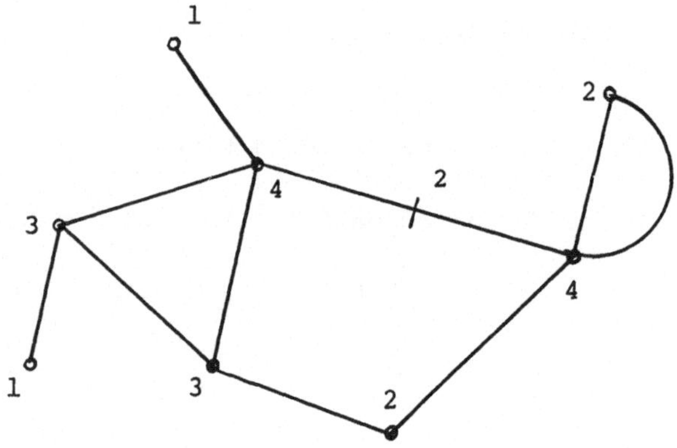

Figure 8.

two nonend points other than x. The order $\omega(x)$ of x is the number of these "cut points". At a non-node the order is always 2. The order is a topological invariant in this sense. If G,G' are homeomorphic graphs and x of G, and x' of G' are corresponding points under the homeomorphism then $\omega(x)$ in $G = \omega(x')$ in G'.

Note this property: the number N_k of points of G whose order is k is a topological invariant of G.

In Figure 8, the numbers indicate orders of the points.

<u>Arcs, loops, trees</u>. An arc of G is merely a polygonal line made up of closed branches. In other words it is a closed arc in the ordinary sense of the term. For want of a more suitable name, "arc" in graph theory will always have the meaning just described. Schematically, one may represent it as $n_1', n_2', \ldots, n_s'$ where the nodes are all distinct and consecutive terms n_h', n_{h+1}' are endpoints of a branch. We will also say: n_1' and n_s' are joined by an arc in G.

A loop of G is a closed polygon line, that is a Jordan curve made up of closed branches. Natural designation: $n_1' n_2', \ldots, n_s' n_1'$ where the n_h' are all distinct and n_h' and n_{h+1}' are again endpoints of a branch.

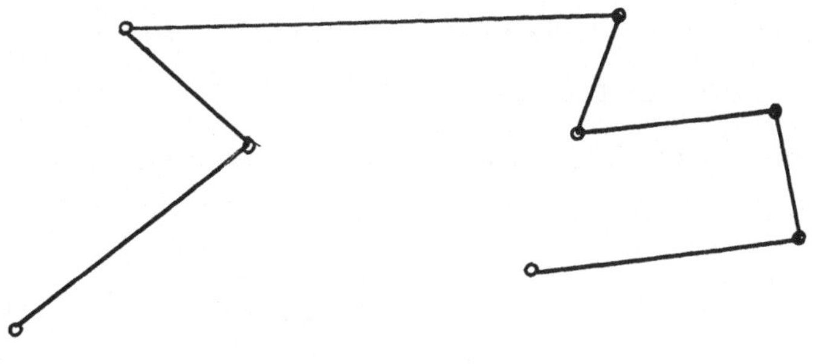

Figure 9.
Arc of G

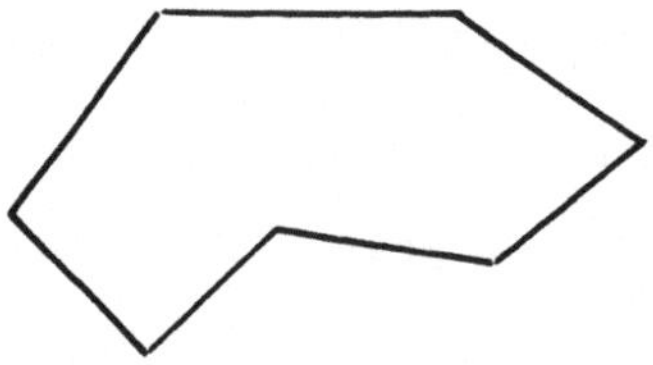

Figure 10 -- Loop

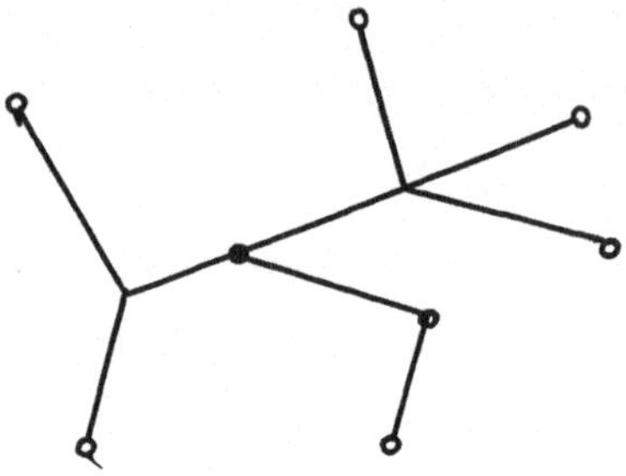

Figure 11 -- Tree

A <u>tree</u> is a connected graph without loops. A finite collection of disjoint trees is a <u>forest</u>.

<u>Polyhedron of G, its components and those of G.</u> The polyhedron of G, denoted by $|G|$ is the collection of all the points in the closed branches of G. One may think of G as a superposed organization over the polyhedron, and one says also: G <u>covers</u> $|G|$.

Let n_1 be any node of G. The set of all branches whose terminal nodes can be joined to n_1 by an arc makes up a subgraph G_1 of G. Similarly for $G - G_1$, etc. One obtains thus a collection of disjoint subgraphs $G_1, G_2, \ldots, G_s$, the <u>components</u> of G. It is immediate that the polyhedral $|G_h|$ are the components of $|G|$ in the sense of Chapter III, Section 3.

Since a tree $\mathscr{C}$ is a connected graph any two nodes of $\mathscr{C}$ may be joined in $\mathscr{C}$ by an arc λ. Interesting enough

(1.1) <u>The arc λ joining any two nodes $n_i^!, n_j^!$ of the tree is unique</u>.

For suppose that there are two such arcs

$$\lambda' = n_i^! n_j^! \ldots n_s^! n_j^! \ , \quad \lambda'' = n_i^! n_1'' \ldots n_t'' n_j^! \ .$$

Let n'_k be the first n' which is also an n''_1 say $n'_k = n''_h$. Then $\mu = n'_i n'_j \ldots n'_k n''_{h-1} n''_{h-2} \ldots n''_1 n'_i$ is a loop in $\mathscr{C}$. Since this is ruled out (1.1) follows

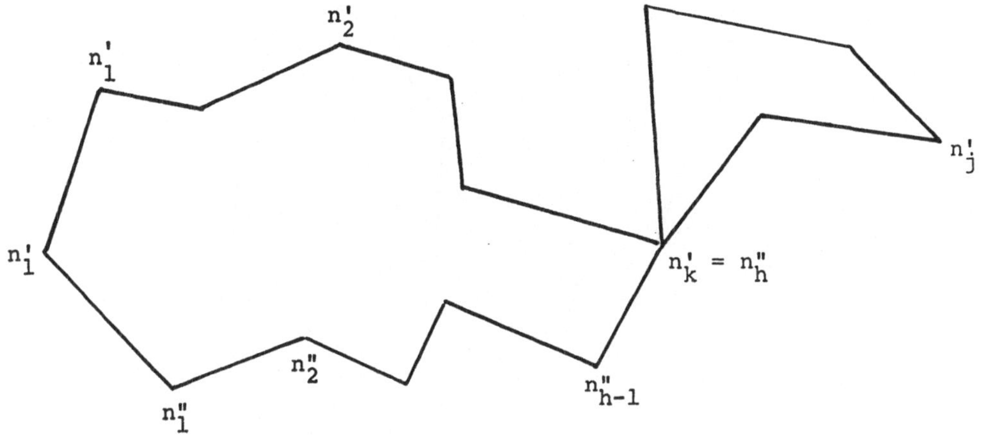

Figure 12.

(1.2) A tree always has a node of order unity. Let n'_1 be any node of the tree $\mathscr{C}$ and let its order $\omega(n'_1) > 1$ let b'_1 be a branch ending at n'_1. Let n'_2 be the other endpoint of b'_1. If the order $\omega(n'_2) > 1$ there is a second branch b'_2 ending at n'_2. Let n'_3 be the other endpoint of b'_2, etc. There results a sequence of nodes $n'_1 n'_2 n'_3 \ldots$, in which n'_h and n'_{h+1} are joined by b'_h. In the sequence all nodes are distinct since otherwise a loop of $\mathscr{C}$ would appear. Since the sequence is finite it must end, and it can only do so at a node of order unity.

2. Subdivision. Characteristic Betti Number

Let the branch b_j of G join the two nodes n_h, n_k. Take a point n' of b_j and replace b_j by the union: arc $n_h n' + n' +$ arc $n' n_k$. This replaces G by a new graph G_1 such that $|G| = |G_1|$. This operation is <u>elementary subdivision</u> and its repetition is <u>subdivision</u>.

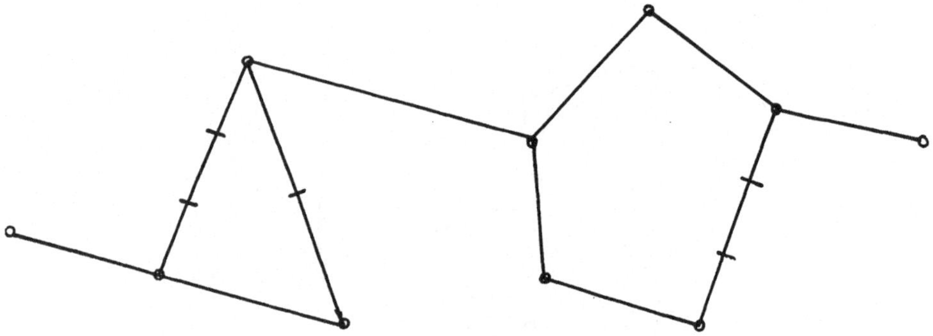

Figure 13.

Take now two graphs G, G' whose polyhedra $|G|$ and $|G'|$ are homeomorphic. If one inserts in G the images of the nodes of G' and in G' those of the nodes of G one obtains two graphs G_1, G_1' with identical disposition of nodes and branches. We may, therefore, consider G_1 and G_1' as identical. Thus G <u>and</u> G' <u>have identical subdivisions</u>. Hence to show that a given property is topologically invariant it is sufficient to show that it is unchanged under an elementary subdivision.

As a first application one may show immediately that:

(2.1) <u>Arcs, loops, and trees are topologically invariant</u>.

(2.2) <u>The number</u> ρ <u>of components of a graph is a topological invariant</u>.

<u>Characteristic</u>. It is by definition the number

$$\chi(G) = \alpha_0 - \alpha_1$$

(number of nodes - number of branches).

(2.3) $\chi(G)$ <u>is a topological invariant</u>.

For our elementary subdivision increases both α_0 and α_1 by unity and so does not affect $\alpha_0 - \alpha_1$.

(2.4) <u>The characteristic of a loop is zero, that of an arc or of a tree is unity</u>.

For a loop or an arc the proof is immediate. Let $\mathscr{C}$ be a tree with α_0 nodes and α_1 branches. Let n_1' be a node of order unity (see 1.2). Let b_1' be the unique branch ending at n_1'. Upon removing $b_1' + n_1'$ what is left is still a tree $\mathscr{C}'$ and $\chi(\mathscr{C}) = \chi(\mathscr{C}')$. By repeating the process one arrives at a single point P and $\chi(\mathscr{C}) = \chi(P) = 1$.

Suppose now that G is connected and let $\mathscr{C}_m$ be a maximal tree of G that is one which ceases to be a tree if augmented by a single branch.

(2.5) $\mathscr{C}_m$ <u>includes all the nodes of</u> G.

Suppose that there is a node n_1' in $G - \mathscr{C}_m$. Let n_k' be a G node of $\mathscr{C}_m$. Since G is connected, these two nodes may be joined by an arc $\lambda = n_1',\ldots,n_s'n_k'$. In the sequence $n_1',\ldots,$ there is a first one say n_h' which is in $\mathscr{C}_m$ and $n_h' \neq n_1'$. Hence the arc $\lambda = n_1'\ldots n_h'$ has only n_h' in $\mathscr{C}_m$. Therefore the addition of the arc μ to $\mathscr{C}_m$ does increase it and $\mathscr{C}_m + \mu$ is a tree $\mathscr{C}_m'$

and larger than $\mathscr{C}_m$. Since this contradicts the maximality of $\mathscr{C}_m$ our assertion follows.

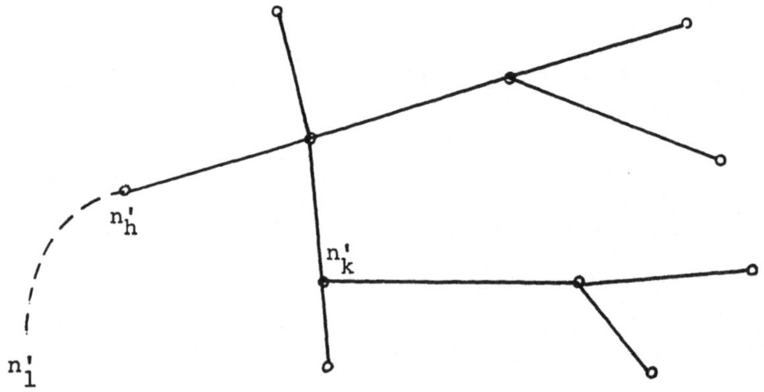

Figure 14.

We conclude then that $G - \mathscr{C}_m$ consists of a certain number R of branches $\bar{b}_1, \bar{b}_2, \ldots, \bar{b}_R$. The number R is the Betti number of G.

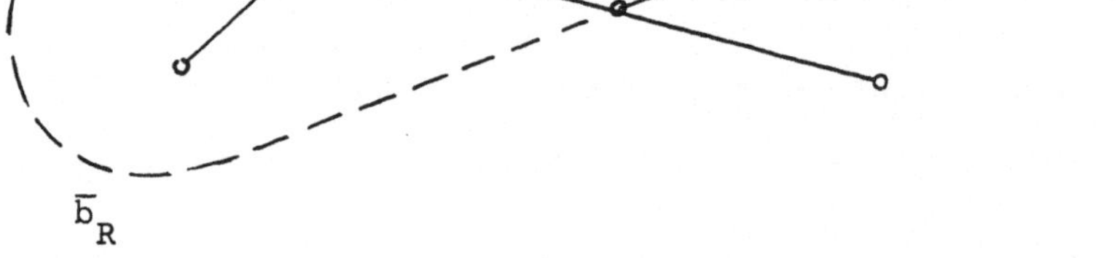

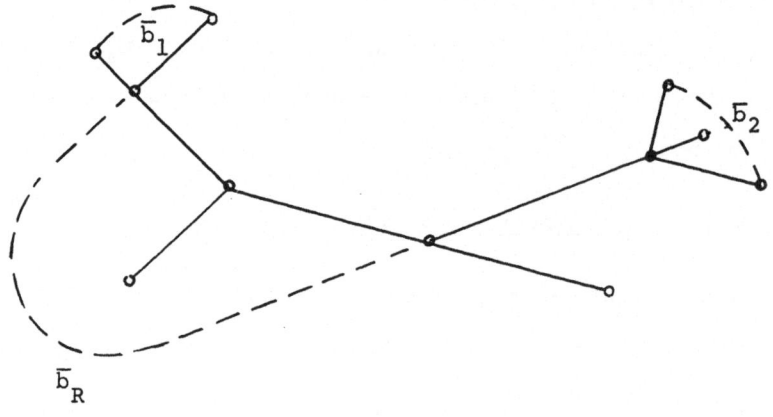

Figure 15.

(It is actually the first Betti number of topology, written R_1, but since we only consider one we prefer the simpler notation R.)

We have then $\chi(G) = \chi(\mathscr{C}_m) - R = 1 - R$.

Explicitly.

(2.6) The characteristic of the graph G is $1 - R$.

Since $\chi(G)$ is a topological invariant we also have

(2.7) $R(G)$ is a topological invariant of the connected graph.

It may be observed that the maximal tree $\mathscr{C}_m$ need not be unique. For if say G is a loop of $n \geq 3$ branches a maximal tree is the arc left when one branch is removed. Hence in this case there are n possible maximal trees.

Let n_j, n_k be the terminal nodes of the branch $\bar{b}_h$. Since $\mathscr{C}_m$ is connected there is a unique arc μ_h joining n_j, n_k in $\mathscr{C}_m$. Hence $\lambda_k = \mu_h + \bar{b}_h$ is a loop, also written $\lambda(\bar{b}_h)$ determined by $\bar{b}_h$. This loop is the only one containing the branch $\bar{b}_h$.

General graph. Let G consist of ρ components $G_1, G_2, \ldots, G_\rho$ and set

$$R(G_h) = R_h , \quad R(G) = R = \sum R_h .$$

Then

$$\chi(G) = \sum \chi(G_h) = \sum (1-R_h) = \rho - R. \qquad (2.8)$$

Since $\chi(G)$ and ρ are topological invariants this formula proves:

(2.9) The Betti number R of a general graph is a topological invariant.

Let $\mathscr{C}_m^h$ be a maximal tree of the component G_h and let $\mathscr{C}_m$ denote the maximal forest $\sum \mathscr{C}_m^h$. Since $G_h - \mathscr{C}_m^h$ consists of branches joining pairs of nodes of $\mathscr{C}_m^h$ we have

(2.10) $G - \mathscr{C}_m$ <u>consists of branches</u> $\bar{b}_1, \ldots, \bar{b}_R$.

CHAPTER V

GRAPH ALGEBRA

1. Preliminaries

Orientation. The first step in dealing with graph algebra is to orient the graph G. This means assigning to each branch b_j with endpoints n_h, n_k one of the two, say, n_h as initial point and the other n_k as terminal point. Hereafter G will always be assumed oriented.

Notice that if one reverses the orientation of b_j by interchainging the roles of n_h and n_k, the branch with its new orientation will be designated by $(-b_j)$.

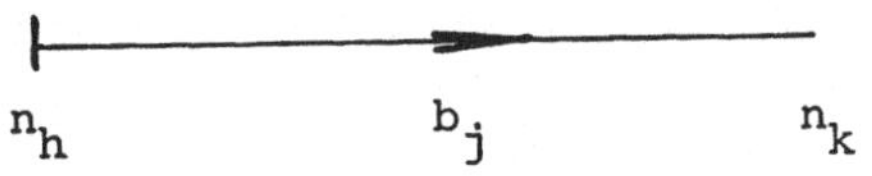

Oriented Branch

Figure 16.

Schematically the process of orienting the branch b_j may be represented by placing a pointed arrow on its symbol.

Chains, cycles and their spaces. Our algebra rests upon the interrelations of two vector spaces C_0 and C_1 whose elements: zero- and one-chains are defined by formal expressions

$$\underline{c}_0 = \sum x_h n_h = \underline{x}'\underline{n}, \quad \underline{c}_1 = \sum y_k b_k = \underline{y}'\underline{b}$$

where the coefficients x_h, y_k are real. The convention is made that $y_k(-b_k) = -y_k b_k$.

Corresponding to the orienting process described for b_j write a formal <u>boundary relation</u>

$$\delta b_j = n_k - n_h .$$

This relation is conveniently set in the general form

$$\delta b_j = \sum \eta_{js} n_s$$

where $\eta_{jk} = 1$, $\eta_{jh} = -1$ and otherwise $\eta_{js} = 0$. The number η_{js} is the <u>incidence number</u> of the branch b_j with the node n_s and $\eta = [\eta_{js}]$ is the <u>incidence matrix</u> of the graph.

Once δ is defined for each branch, that is for the elements of a base for the space C_1 one can extend it to all of C_1 as a linear transformation $\delta\colon C_1 \to C_0$ by the relation

$$\delta \underline{c}_1 = \delta \sum y_j b_j = \sum y_j \delta b_j = \sum y_j \eta_{jk} n_k$$

or in vector form

$$\delta \underline{y}'\underline{b} = \underline{y}'\eta\underline{n} .$$

One may define a similar operation, still written $\delta\colon C_0 \to 0$ (since there is no C_{-1}). This means that zero-chains have no boundary $\neq 0$.

Let Z_1 be the nucleus of the transformation $\delta\colon C_1 \to C_0$. Its elements $\underline{z}$ are the one-chains without boundary that is such that $\underline{z}'\eta\underline{n} = 0$. Since the n_h satisfy no relation, the equation just written is equivalent

$$\underline{z}'\eta = 0. \tag{1.1}$$

This is the characteristic equation of one-cycles. Since $\delta C_0 = 0$ all zero-chains are cycles.

The simplest type of one-cycle is one attached to a loop λ. Let $\lambda = n_1'n_2',\ldots,n_s'n_1'$ where the successive branches $b_1'b_2',\ldots,$ are so oriented that $\delta b_h' = n_{h+1}' - n_h'$ $(n_{s+1}' = n_1')$. Since $\delta\,\Sigma\, b_h' = 0$, $\Sigma\, b_h'$ is a one-cycle and this is the one that we had in mind. For convenience we keep the designation λ for this cycle. Note that the full orientation of λ is determined by that of any b_h'. We may also refer to the cycle as $-\lambda(-b_1')$ and we have $\lambda(b_1') = \lambda(b_2') = \ldots = -\lambda(-b_1') = -\lambda(-b_2') = \ldots = n$.

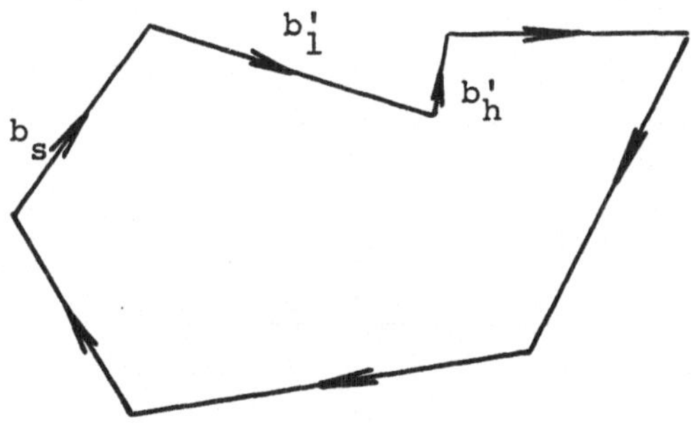

Figure 17.

Since Z_1 is the nucleus of $\delta\colon C_1 \to C_0$ we have

$$C_1 = D_1 \oplus Z_1$$

where D_1 is the space of the one-chains with non-zero boundary (except $d_1 = 0$).

Since $C_0 = Z_0$, the analogue of D_1 for the space C_0 is $D_0 = 0$.

Let $\delta D_1 = F_0$. The space F_0 consists of zero-chains which are boundaries of one-chains. Since δ has no nucleus as an operation $D_1 \to F_0$, the two spaces D_1 and F_0 are isomorphic. We also have

$$C_0 = Z_0 \doteq F_0 \oplus H_0 \doteq D_1 \oplus H_0 \tag{1.2}$$

where $H_0 = C_0/F_0$ (recall that according to Chapter II, Section 5 "$\cdot$" signifies that subspaces have been replaced by isomorphs).

2. Dimensional Calculations

For the two dimensions 0,1 we have defined the spaces C_i, Z_i, D_i, F_i, H_i, $i = 0,1$, some of them being zero. To within isomorphisms their mutual relations are represented schematically in Figure 18

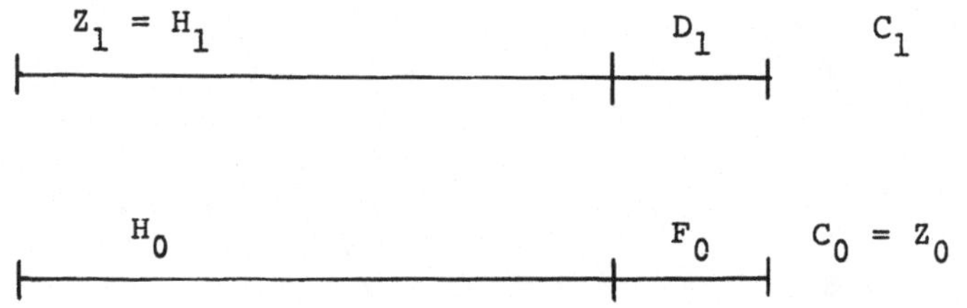

Figure 18.

with the zero terms not represented.

Since $\dim C_0 = \alpha_0$, $\dim C_1 = \alpha_1$, one merely needs to find the dimension of one of the two terms in each segment.

We begin with $\dim H_1$. Let $\mathscr{C}_m$ and the $\bar{b}_j$ have the same meaning as in Chapter IV, Section 2, except that now $\bar{b}_j$ is oriented. Let $\lambda(\bar{b}_j) = \lambda_j$ be the associated one-cycle. We have then

$$\lambda_j = \bar{b}_j + \mu_j$$

where μ_j is an oriented arc in $\mathscr{C}_m$.

If $\underline{c}_1$ is any one-chain let $[c_1]$ stand for the graph which is the union of the closed branches actually present in the expression of $\underline{c}_1$.

(2.1) <u>The subgraph</u> $[z_1]$ <u>of a cycle</u> $\underline{z}_1$ <u>always contains a loop</u>.

Suppose that

$$\underline{z}_1 = \alpha_1 b_1 + \alpha_2 b_2 + \cdots$$

and suppose the b's so numbered that $\alpha_1 \neq 0$. Let $\delta b_1 = n_2' - n_1'$. Since $\delta \underline{z}_1 = 0$ the node n_2' appears in a branch, say $b_2 \neq b_1$ with a coefficient $\alpha_2 \neq 0$, and in such manner that $\delta b_2 = n_3' - n_2'$, etc. One obtains thus an arc $n_1' n_2' \ldots n_s'$ of $[\underline{z}_1]$. Since the process must end sometime $n_t' = n_h'$, $h < t$, so that $n_h', \ldots, n_t'$ is a loop contained in $[\underline{z}_1]$.

Since the forest $\mathscr{L}_m$ contains no loop, we also have

(2.2) <u>A forest contains no 1-cycle</u>.

Denote in general by $\underline{f}$ any one-chain of $\mathscr{L}_m$ (maximal forest of the graph G). From (2.2) there follows:

(2.3) <u>If</u> $\underline{f}$ <u>is a cycle then</u> $\underline{f} = 0$.

Write now

$$\lambda_k = \bar{b}_k + \underline{f}, \quad \underline{z}_1 = \sum \beta_k \bar{b}_k + \underline{f}.$$

This implies that

$$\underline{z}_1 - \sum \beta_k \lambda_k = \underline{f} = 0$$

since the first term is a cycle. <u>Thus every one-cycle depends upon the cycles</u> λ_k.

On the other hand the λ_h are independent. For a relation

$$\sum \gamma_h \lambda_h = 0$$

yields

$$\sum \gamma_k \bar{b}_h = f.$$

However, f includes no $\bar{b}_k$, and so this relation implies that every $\gamma_k = 0$.

Conclusion: $\{\lambda_h\}$ is a base for the one-cycles. Hence

$$\dim H_1 = \dim Z_1 = R_1 = R. \tag{2.4}$$

From the sketch (Figure 18) one infers then that

$$\dim D_1 = \alpha_1 - R = \dim F_0 . \tag{2.5}$$

Therefore, from the sketch

$$\dim H_0 = \alpha_0 - \dim F_0 = \alpha_0 - \alpha_1 + R = \rho. \tag{2.6}$$

This completes the calculations of the space dimensions.

Remark. The preceding calculations rest directly upon the geometric evaluation of the numbers R and ρ. One may, however, calculate R immediately in terms of the matrix η. For if its rank is r we have at once for (1.1):

$$\dim Z_1 = \alpha_1 - r = R. \tag{2.7}$$

3. Space Duality. Co-theory

The spaces of chains C_0, C_1 of a graph G and operator δ are an obvious example of the situation cinsidered in Chapter II, Sections 4, 5. There arises then an associated graph duality. The only deviation is the reference to the various elements as cochains, cocycles, etc., and so one speaks of the co-theory.

Corresponding to the nodes n_h and branches b_k of G introduce new co-elements n_h^*, b_k^* and their zero- and one-cochains

$$\underline{c}_0^* = \sum x_h^* n_h^* = \underline{x}^{*\prime}\underline{n}^*, \quad \underline{c}_1 = \sum y_k^* b_k^* = \underline{y}^{*\prime}\underline{b}^*,$$

generating spaces C_0^* and C_1^*.

Reversing the earlier boundary scheme we ask now what branches b_s^* (same as b_s) end at the node n_k^* (same as n_k). The resulting coboundary is

$$\delta^* n_k^* = \sum \eta_{sk} b_s^* ,$$

and hence for a cochain $\Sigma\, y_k^* n_k^*$

$$\delta^* \sum y_k^* n_k^* = \sum y_k^* \eta_{sk} b_s^* , \quad \delta^* \underline{y}^{*\prime}\underline{n}^* = \underline{y}^{*\prime}\eta^\prime \underline{b}^*.$$

This shows that the matrix of δ^* is η'. Thus C_1, C_0, δ are related to C_1^*, C_0^*, δ^* like A, B, ϕ to A^*, B^*, ϕ^* in Chapter II, Section 5. All that is required, therefore, is to adapt the notations to the present situation. We shall not attempt to do so in full, but will only obtain the explicit form of a property of importance later.

Referring then to Chapter II, (5.11) we find

(3.1) <u>A n.a.s.c. in order that $\underline{z}_1^\prime\underline{b}$ be a one-cycle is that, whatever the coboundary $\underline{u}_1^{*\prime}\underline{b}^*$ we have</u>

$$\underline{z}_1^\prime \underline{u}_1^* = 0. \tag{3.2}$$

Let the spaces C_1 and C_1^* be referred to the same coordinates. The preceding property affirms then

(3.3) <u>The subspaces Z_1 of the one-cycles and F_1^* of the coboundaries are orthogonal and complementary, that is</u>

$$C_1 \doteq Z_1 \oplus F_1^* .$$

(3.4) A direct proof of (3.3). A one-cycle is any one-chain $\underline{z}'\underline{b}$ such that $\delta\underline{z}'\underline{b} = \underline{z}'\eta\underline{n} = 0$ which yields $\underline{z}'\eta = 0$. The coboundary of a cochain $\underline{y}^{*\prime}n^*$ is $\underline{y}^{*\prime}\eta'\underline{b}^*$, that is it is a cochain $\underline{u}^* = \eta\underline{v}^*$. Hence

$$\underline{z}'\underline{u}^* = \underline{z}'\eta\underline{v}^* = 0.$$

Property (3.3) has an important application to network theory.

Dimensional calculations for the dual spaces. We first draw a sketch analogous to Figure 18.

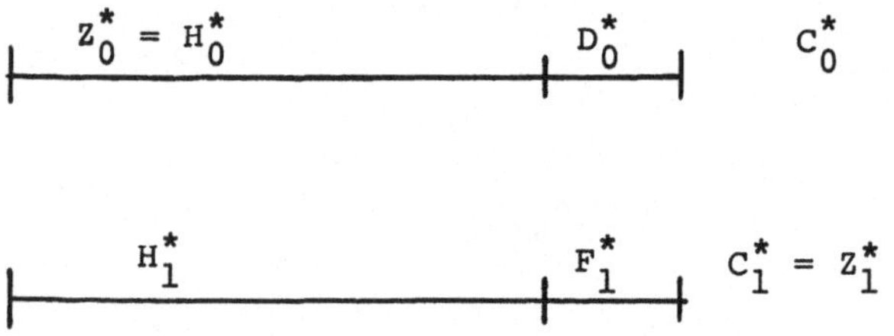

Figure 19.

It implies

$$\dim C_0^* = \alpha_0, \quad \dim C_1^* = \alpha_1$$

Also from (3.3): $\dim F_1^* = \alpha_1 - R = \dim D_1$. Hence

$$\dim H_0^* = \alpha_0 - (\alpha_1 - R) = \rho = \dim H_0;$$

$$\dim H_1^* = \alpha_1 - (\alpha_1 - R) = R = \dim H_1.$$

That is:

(3.5) The cohomology spaces have the same dimension as the corresponding homology spaces.

CHAPTER VI

ELECTRICAL NETWORKS

An electrical network is just a graph to whose branches there are assigned two functions: a voltage distribution and a current distribution, each subjected to classical laws due to Kirchoff. We first discuss these laws and reformulate them in keeping with the topological development of the preceding chapter.

1. Kirchoff's Laws

Let N be a network and G its graph. All notations of the previous chapters will be applied to G.

To each (oriented) branch b_h of G there are assigned a voltage v_h and a current i_h positive in the same directions as b_h itself, and making up vectors $\underline{v}$ and $\underline{i}$. For the present they are merely subjected to Kirchoff's laws. We state them and reformulate them.

(1.1) First Kirchoff law (current law). A current distribution is any vector $\underline{i}'\underline{b}$, such that the algebraic sum of the currents arriving at any node is zero.

The physical meaning of this law is that a node is neither a source nor a sink of electrical fluid.

Let n_k be an endpoint of b_h. Then i_h arrives at or leaves n_k accordingly as $i_h\eta_{hk} > 0$ or < 0.

The obvious conclusion is that the algebraic sum of the currents arriving at the node n_k is

$$\sum_h i_h\eta_{hk} .$$

Hence Kirchoff's first law means that $\underline{i}'\eta = 0$ or that the chain $\underline{i}'\underline{b}$ is a cycle. The reformulation of the law is therefore:

(1.2) A current distribution is any vector $\underline{i}'\underline{b}$ which is a cycle of the graph G.

(1.3) Second Kirchoff law (voltage law). A voltage distribution is any cochain vector $\underline{v}'\underline{b}^*$ such that the algebraic sum of the voltages along any loop is zero.

Reformulation of the second law. One may state the second law in this way. Let any loop be given by

$$\lambda = \sum \varepsilon_k b_k , \quad \varepsilon_k = \pm 1 \text{ or } 0.$$

Then

$$\sum \varepsilon_k v_k = 0.$$

In particular let $\lambda_1, \lambda_2, \ldots, \lambda_R$ be a maximal set of independent loops and set

$$\lambda_j = \sum \varepsilon_{jk} b_k .$$

If v is the same as above then

$$\sum_k \varepsilon_{jk} v_k = 0, \quad j = 1, 2, \ldots, R.$$

Let $i^{(h)}$ denote the current distribution represented by the cycle $\alpha_h \lambda_h$, its value being α_h in each branch of λ_h. We have then

$$\alpha_h \sum \varepsilon_{hk} v_k = 0 = \underline{i}^{(h)'}\underline{v}.$$

Since every current distribution $\underline{i}$ depends upon the $\underline{i}^{(h)}$ we have

$$\underline{i}'\underline{v} = 0. \tag{1.4}$$

That is the vector $\underline{v}$ of the voltage distribution is orthogonal to every current vector. Therefore v is a coboundary.

Conversely, if v is a coboundary it will be orthogonal to the special current λ and the second law will be obeyed. Hence this reformulation of the second law:

A voltage distribution is any coboundary (of a zero-cochain). That is $\underline{v}'\underline{b}^* = \delta^*\underline{w}'\underline{n}^* = \underline{w}'\eta'\underline{n}^*$ or finally $\underline{v}' = \underline{w}'\eta'$, hence

$$\underline{v} = \eta\underline{w}. \tag{1.5}$$

Thus if $\delta b_h = n_k - n_j$ then $v_h = w_k - w_j$. The w's are the electrostatic potentials and are arbitrary. We may therefore state:

(1.6) One may assign arbitrary electrostatic potentials w_s and then determine the voltages v_j of the branches, by the following rule: If $\delta b_h = n_k - n_j$ then $v_h = w_k - w_j$.

Returning to (1.4) notice that since $\underline{v}$ is a vector in a linear subspace, the latter (that is $V_{\alpha_1 - R}$) likewise contains $d\underline{v}$. Hence from (1.4)

$$\underline{i}'d\underline{v} = 0 = d\underline{i}'\underline{v} \tag{1.7}$$

a relation of much importance in the rest of the chapter.

2. Different Types of Elements in the Branches

One may envisage various types of electrical mechanisms in the branches. Through the insertion of new nodes which does not modify the situation in any manner one may always assume that each branch contains a single mechanism such as we will now envisage.

We distinguish then the following typical branches:

I. Resistor. Basically this is the name given to any branch b in which a current i and voltage v are related by a single relation $F(i,v) = 0$. We shall always assume that F is defined for all real values of i and v, that it has continuous first partials F_i, F_v throughout the plane i,v, and moreover that F_i, F_v never vanish simultaneously. Geometrically it means that the curve $F = 0$ has a continuously turning well defined tangent and no multiple points (or other singularities).

The simplest instance is the <u>linear</u> (ohmic) resistance $v = Ri$, (R a positive constant). Of interest also are a voltage or constant current generator, when $F = 0$ is respectively a line parallel to the i axis, and one parallel to the v axis. Still another noteworthy case is the vacuum tube. Van der Pol had proposed the simple characteristic

$$v = ai - bi^3,$$

a and $b > 0$. A more realistic description is a characteristic

$$v = ai - bi^3 + ci^5, \quad a,b,c > 0.$$

II. <u>Inductor</u>. Here the relation between i and v is

$$L(i)\,\frac{di}{dt} = v, \quad L(i) \geq 0, \tag{2.1}$$

where we accept for $L(i)$ the same general type as for $R(i)$. This is the case of the potential induced in a solenoid or electromagnet by a current $i(t)$.

III. <u>Capacitor</u>. Same situation with i,v interchanged, the relation now being

$$C(v)\,\frac{dv}{dt} = i, \quad C(v) \geq 0.$$

3. A Structural Property

By means of a maximal forest $\mathscr{C}_m$ of the graph G the following property has been obtained: Let $\bar{b}_1, \bar{b}_2, \ldots, \bar{b}_R$ be the branches of $G - \mathscr{L}_m$. Let f designate any chain of $\mathscr{L}_m$. Then there exist R linearly independent loops $\lambda_1,\ldots,\lambda_R$ such that

$$\lambda_h = \bar{b}_h + f_h. \tag{3.1}$$

We propose to describe an algebraic method for arriving at the same result. To that end let $\lambda_1,\lambda_2,\ldots,\lambda_R$ be a base for cycles and let

$$\gamma_h = \sum \varepsilon_{hk} b_k.$$

Since the γ's are independent the matrix $[\varepsilon_{hk}]$ is of rank R. Let the branches be so ordered that the matrix $D = [\varepsilon_{hk}]$; $h,k = 1,2,\ldots,R$, is already nonsingular. Upon replacing the vector

$$\underline{\gamma} = \begin{bmatrix} \gamma_1 \\ \gamma_2 \\ \cdots \\ \cdots \\ \gamma_R \end{bmatrix}$$

by $D^{-1}\gamma$ there will result a set of cycles, still called $\gamma_1,\ldots,\gamma_R$ such that the corresponding $D = 1$.

Now if a branch is in no γ_h it will receive no current and hence it is electrically <u>inactive</u>. We may then as well surpress it. In other words we merely replace the graph G by the subgraph $[\gamma_1] + [\gamma_2] + \cdots + [\gamma_R]$. To simplify matters assume that this graph is G itself.

Let H be the subgraph composed of the closed branches $b_{R+1},\ldots,b_{\alpha_1}$. There is no cycle in H, since such a cycle would have to be independent of the γ_h. Thus H is a <u>forest</u>.

Let now λ_h be a loop contained in $[\gamma_h]$. Since λ_h is not a chain f we have (replacing perhaps λ_h by $-\lambda_h$)

$$\lambda_h = b_h + f, \quad \gamma_h = b_h + f' .$$

Hence $\lambda_h - \gamma_h = f - f' = 0$, since $f - f'$ is a cycle. Hence $\gamma_h = \lambda_h$. Thus we have reproduced by an algebraic attack the exact situation announced at the beginning of the section.

Since λ_h are loops we have

$$\lambda_h = b_h + \sum_s \varepsilon_{hs} b_{R+s} \tag{3.2}$$

$$\varepsilon_{hs} = \pm 1 \quad \text{or} \quad 0.$$

4. Differential Equations of an Electrical Network

We propose to assign a certain number p of independent currents and q of independent voltages to express all currents and voltages in terms of these and write the corresponding equations of the network.

Since there are only R independent current distributions we must have $p \leq R$.

With the same notations as in Section 3 choose then $\overline{b}_1, \ldots, \overline{b}_p$ as inductor branches and let any one of these be designated by b^*_μ and their currents by i^*_μ .

Let the branches $\overline{b}_{p+1}, \ldots, \overline{b}_R$ be resistors and let $\overline{b}_\rho$ denote any one of these branches. The loop

$$\lambda_\rho = b_\rho + \sum \varepsilon_{\rho\nu} b^*_\nu$$

where b^*_ν designates any branch of a chain $\lambda_\rho - b_\rho$. The branches b^*_ν are taken as capacitors and assigned voltages v^*_ν ; their number is q and their currents are i_ν .

The remaining branches (if any): $b_{R+q+1}, \ldots, b_{\alpha_1}$ are denoted by b_σ and they are again resistors.

(4.1) The current and voltage distribution $\underline{i},\underline{v}$ depend entirely upon the independent variables i_μ^* and v_ν^*.

By the second law v_ρ is a sum of voltages v^* :

$$-v_\rho = \sum \varepsilon_{\rho\nu} v_\nu^* . \tag{4.2}$$

Since b_ρ is a resistor

$$i_\rho = g_\rho(v_\rho) \tag{4.3}$$

and hence the current in every branch is determined by the currents i_μ^* and i_ρ (first law). Hence they are functions of the i_μ^* and v_ν^*. Thus in every branch b_j either v_j is given (one of the v_ν^*) or else determined by the resistor relations, or else again if a b_μ^* from the second law (in λ_μ).

Our next step will result from the application of the relation

$$\underline{v}'d\underline{i} = 0. \tag{4.4}$$

Upon taking account of the relation

$$d(v_\nu^* i_\nu) = v_\nu^* di_\nu + i_\nu dv_\nu^*$$

we obtain

$$\sum v_\mu di_\mu^* - \sum i_\nu dv_\nu^* + \sum v_\rho di_\rho + \sum v_\sigma di_\sigma + \sum d(i_\nu v_\nu^*) = 0. \tag{4.5}$$

We shall show that each of the last three sums is an exact differential. Conventionally we shall write

$$f(i)di = dF(i), \quad g(v)dv = dG(v).$$

Now

$$v_\rho di_\rho = f_\rho(i_\rho)di_\rho = dF_\rho(i_\rho) = dF_\rho(g_\rho(v_\rho)) = dH_\rho(v^*).$$

Hence

$$\sum v_\rho di_\rho = dH(v^*), \quad H = \sum H_\rho . \tag{4.6}$$

Next, since b_σ is a branch of one or more λ_μ we have

$$i_\sigma = \sum_\mu \varepsilon_{\mu\sigma} i^*_\mu , \quad v_\sigma = f_\sigma(i_\sigma).$$

Hence

$$v_\sigma di_\sigma = f_\sigma(i_\sigma) di_\sigma = dF_\sigma(i_\sigma) = dK_\sigma(i^*)$$

and, therefore

$$\sum v_\sigma di_\sigma = dK(i^*), \quad K = \sum K_\sigma . \tag{4.7}$$

Finally,

$$\begin{aligned} i_\nu &= \sum_\mu \varepsilon_{\mu\nu} i^*_\mu + \sum_\rho \varepsilon_{\rho\nu} i_\rho \\ &= \sum_\mu \varepsilon_{\mu\nu} i^*_\mu - \sum_\rho \varepsilon_{\rho\nu} g_\rho \left(\sum \varepsilon_{\rho\nu} v^*_\nu\right). \end{aligned}$$

Hence

$$\sum i_\nu v^*_\nu = \sum \varepsilon_{\mu\nu} i^*_\mu v^*_\nu + L(v^*). \tag{4.8}$$

Hence, if we set $M = H + L$ then $-P(i^*,v^*) = K(i^*) + M(v^*) + i^{*\prime}\varepsilon v^*$ we find from (4.5)

$$\sum v_\mu di^*_\mu - \sum i_\nu dv^*_\nu = dP(i^*,v^*). \tag{4.9}$$

Since i^*_μ , v^*_ν are independent variables upon equating the coefficients of their differentials we obtain

$$\frac{\partial P}{\partial i^*_\mu} = v_\mu , \quad -\frac{\partial P}{\partial v^*_\nu} = i_\nu . \tag{4.10}$$

Hence, the final "potential system"

$$\begin{aligned} L_\mu(i_\mu^*)\dot{i}_\mu^* &= \frac{\partial P}{\partial i_\mu^*} = -\frac{\partial K}{\partial i_\mu^*} - \sum_\nu \varepsilon_{\mu\nu} v_\nu^* \\ C_\nu(v_\nu^*)\dot{v}_\nu^* &= -\frac{\partial P}{\partial v_\nu^*} = \frac{\partial M}{\partial v_\nu^*} + \sum_\mu i_\mu^* \varepsilon_{\mu\nu} . \end{aligned} \tag{4.11}$$

Introduce the two matrices

$$L(\underline{i}^*) = \operatorname{diag}(L_1(i_1^*),\ldots,L_p(i_p^*))$$

$$C(\underline{v}^*) = \operatorname{diag}(C_1(v_1^*),\ldots,C_q(v_q^*)) .$$

In terms of these matrices the system (4.11) assumes the condensed form

$$L(i^*)\dot{i}^* = \frac{\partial P}{\partial i^*} , \quad C(v^*)\dot{v}^* = -\frac{\partial P}{\partial v^*} \tag{4.12}$$

where the partials are now gradients.

In our treatment mutual inductance has been neglected. Upon taking it into account L must be replaced by a symmetric (not diagonal) matrix $M(i^*)$ but otherwise (4.12) is unchanged.

CHAPTER VII

COMPLEXES

Our further progress rests upon another excursion into topology: theory of complexes and related polyhedra (only for dimension two), with applications to surfaces in the next chapter. However, we do not plan to pursue topology beyond "piecewise linear" arguments. This is done in order to minimize recourse to more delicate arguments which would be imposed by full fledged topology, and which we really do not need.

1. Complexes

Let $\sigma_p = A_0 A_1 \cdots A_p$ be a p-simplex in Euclidean q-space, $q > p$. Recall that if one thinks of the points A_h as vectors then the A_h are assumed linearly independent and σ_p is the set of all real vectors

$$A = x_0 A_0 + \cdots + x_p A_p , \quad 0 < x_h < 1, \quad \sum x_h = 1.$$

The A_h are the vertices of σ_p, the $A_h A_k$, $h \neq k$, and $A_h A_k A_\ell$, $h \neq k \neq \ell$, are its edges and triangles. It is convenient to think of both as open, that is exclusion of the endpoints of edges, and of perimeters of triangles. The linear independence of the A_h has the following important consequences.

(1.1) Distinct edges or triangles are disjoint.

Let G be our usual graph with its nodes n_h, and branches b_k and let the simplex σ_p be so chosen that $p > \alpha_0$. Take α_0 vertices of σ_p, say $A_1, A_2, \ldots, A_{\alpha_0}$ and for each branch b_h with terminal nodes n_j, n_k draw the edge $A_j A_k$. As a consequence these edges and vertices make up a complete and faithful representation of the graph G with rectilinear branches. In order not to diversify notations to excess we may assume that this representation is G itself. Thus the node n_h is now also the vertex A_h of σ_p and the branch $n_j n_k$ is the edge $A_j A_k$ of the simplex.

Consider now any set of distinct loops $\lambda_1, \lambda_2, \ldots, \lambda_s$ of the graph and suppose that $p \geq \alpha_0 + s$. To the loop λ_h assign the vertex A_{α_0+h} of σ_p. The loop λ_h is now a polygon of σ_p. Draw in σ_p a segment from A_{α_0+h} to every point of λ_h. Let $\lambda_h = n'_1 n'_2 \ldots n'_r n'_1$. The collection of all the segments to the points of $n'_h n'_{h+1}$ $(n'_{r+1} = n'_1)$ is a closed triangle with vertex at A_{α_0+h} . The collection of all these triangles has the same structure as say the collection of triangles, with vertex at the center of a regular r-gon and bases on its sides. This is manifestly the topological image of a circle - that is a closed 2-cell and will be called here "cell" for short. The open cell includes the point A_{α_0+h} but not the base λ_h.

Upon applying the preceding construction to each of the loops λ_h there result s new cells $e_1, \ldots, e_s$ all disjoint from one another and from the graph G itself. The collection K of G plus these s cells is known as a 2-complex, or simply complex for short.

Let $\beta_0, \beta_1, \beta_2$ be the number of nodes, sides and triangles of K. The characteristic of K (introduced by Poincaré) is the expression

$$\chi(K) = \beta_0 - \beta_1 + \beta_2. \tag{1.2}$$

It will play an all important role in what follows.

Let us just count the elements: triangles, edges, vertices

which are in the cell e_h, and let their numbers be $\gamma_0, \gamma_1, \gamma_2$. Since there is just the vertex A_{α_0+h} and equal numbers of sides and triangles we find at once

$$\chi(e_h) = \gamma_0 - \gamma_1 + \gamma_2 = 1.$$

Hence, as far as $\chi(K)$ itself is concerned we may merely count each cell as unity. We then have

$$\chi(K) = \chi(G) + s = \alpha_0 - \alpha_1 + s = \rho(G) - R + s.$$

In keeping with our earlier notations it is best to set $s = \alpha_2$ = the number of cells of K. If we think of α_0, α_1 as numbers of "0-cells" and "1-cells" of K we have the consistent notation

$$\chi(K) = \alpha_0 - \alpha_1 + \alpha_2 = \rho(G) - R + \alpha_2 , \tag{1.3}$$

where α_2 is now the number of loops λ_h.

To sum up the characteristic may equally well be calculated directly from the number of cells e_h (that is from the number of loops λ_h).

Polyhedron. The set of all points in the elements: nodes, branches, cells of K is called a polyhedron, and denoted by $|K|$ or also by Π. The complex K is said to cover the polyhedron $|K|$. The distinction between K and $|K|$ is meant to emphasize the fact that the complex is a geometric figure plus a definite structure: decomposition into nodes, branches and cells.

Connectedness. Since every point of $|K|$ is connected by an arc to a point of the graph G the components Π_h of $|K|$ are polyhedra each uniquely determined by a component G_h of G and there is a subcomplex K_h of K which covers Π_h, so that the latter is merely a connected subpolyhedron of $|K|$. The K_h are the components of K and their number ρ is the same as the number of components of the graph G. Evidently ρ is a topological invariant of $|K|$.

2. Subdivision

It was shown in Chapter V, Section 2 that topologically identical graphs have a common subdivision. As a consequence to prove topological invariance for graph properties it was sufficient to prove their invariance under subdivision. This agreeable situation is much more difficult to establish for complexes. However, we do not require the more stringent topological invariance and so we shall merely consider subdivision invariance.

One must first define subdivision. This is done in two steps. We first define elementary subdivision as consisting of one of the following two operations:

(a) Introduction of a single node in a branch, that is elementary subdivision of the graph G (Figure 20). It replaces the

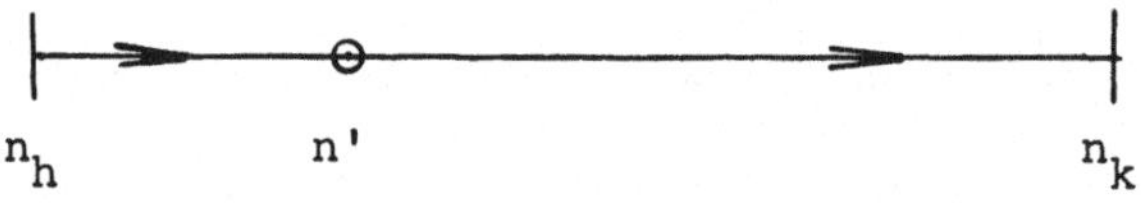

Figure 20.

oriented branch $b_j = n_h n_k$ by the new node n' plus the oriented new branches $n_h n'$ and $n' n_k$.

(b) Introduction in a cell e of a new branch $b' = n_h n_k$ thus replacing e by two new cells e', e'' plus the new branch b'. The various orientations are fully described in Figure 21.

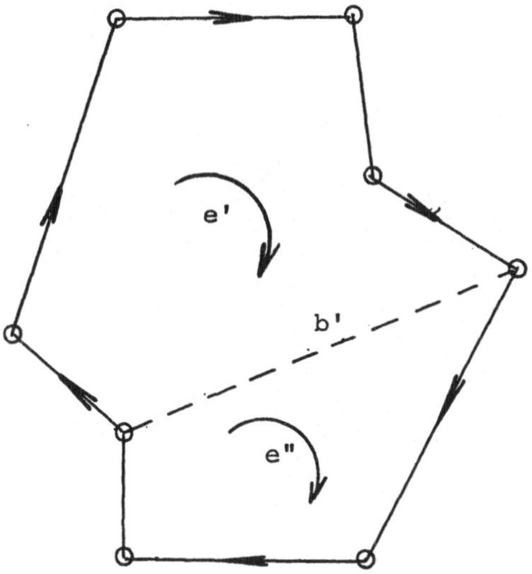

Figure 21.

Note that $\delta e' = -b' + \cdots, \delta e'' = b' + \cdots$ so that $\delta(e'+e'') = \delta e$.

<u>Subdivision</u> itself consists merely of a finite succession of elementary subdivisions.

To prove subdivision invariance it will be sufficient then to prove invariance under elementary subdivision.

There are two noteworthy subdivisions. The first consists merely in joining some point in e by arcs to every vertex of e (Figure 22) thus replacing e by a collection of triangles, edges

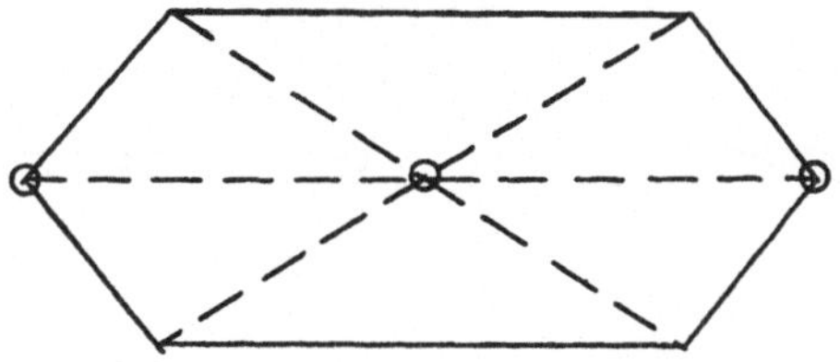

Figure 22.

and a node. The second subdivision consists in joining a point of e by arcs to every node and to a point in each branch of the boundary of e (Figure 23). Generally

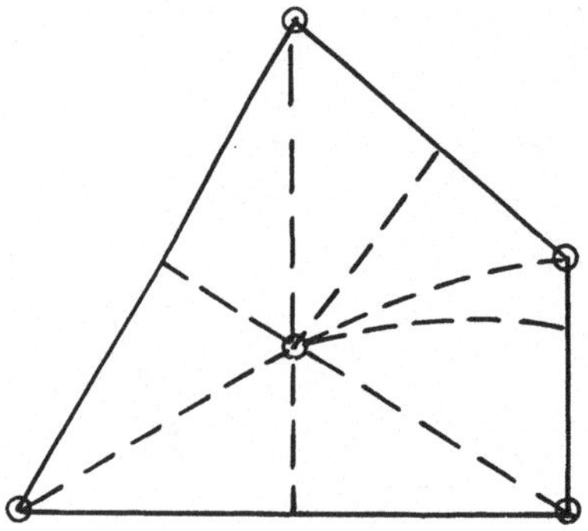

Figure 23.

the new vertices are the centroids of branches and cells.

The operation when applied to the whole complex K is called <u>barycentric subdivision</u>.

Since the number ρ of components of K is actually a topological invariant of the polyhedron $|K|$ it is not affected by subdivision. That is

(2.1) *The number* ρ *of components of the complex* K *is a subdivision invariant.*

Since an elementary subdivision manifestly does not alter $\chi(K)$ we have:

(2.2) *The characteristic* $\chi(K)$ *is a subdivision invariant.*

3. Complex Algebra

It will be obtained by a fairly obvious extension to cells of the notions of chains and their boundaries.

In the next chapter we will have occasion to consider chain, mod 2, that is with coefficients in the field made up of the numbers 0,1 with $0.0 = 0.1 = 1.0 = 0$, $1 + 1 = 0$, $1.1 = 1$. This will merely require to disregard orientations but otherwise nothing will be changed.

Assuming real coefficients, we must first orient the cells e_h. This is done naturally by means of the loops. With each of the two possible orientations of λ_h: $\pm\lambda_h$, we associate a definite orientation of e_h. One may think of it intuitively as orienting e_h by means of a "pointed arrow" along its border. The symbol e_h will refer henceforth to the cell in a definite orientation. The cell with the opposite orientation is denoted by $(-e_h)$.

Let then all the cells be oriented in a definite way. A 2-chain of K is a formal expression with real coefficients x_h

$$c_2 = \Sigma\, x_h\, e_h\ , \tag{3.1}$$

it being understood that $x_h(-e_h) = -x_h\, e_h$.

In our standard vector-matrix notations:

$$c_2 = \underline{x}'\ \underline{e}\ . \tag{3.2}$$

The generation of the vector space C_2 of these 2-chains follows the same route as for 0-chains and one-chains. The dimension of C_2 is α_2.

<u>Boundary operation</u>. We have already defined boundaries and cycles for the graph and its one-chains. For convenience in certain algebraic operations define the boundary operator δ for all three dimensions: 0,1,2. Since there are no (-1)-chains a natural definition for the dimension zero is to set $\delta C_0 = 0$, that is all zero-chains are zero cycles. We define naturally

$$\delta e_h = \lambda_h\ ,$$

hence

$$\delta \Sigma x_h e_h = \Sigma x_h \lambda_h .$$

In terms of the branches b_k we have

$$\lambda_h = \Sigma \zeta_{hk} b_k$$

and hence

$$\delta \Sigma x_h e_h = \Sigma x_h \zeta_{hk} b_k . \tag{3.3}$$

The number (integer) ζ_{hk} is the <u>incidence number from the cell</u> e_h <u>to the branch</u> b_k. It is ± 1 if b_k is a branch of λ_h and zero otherwise. The matrix $\zeta = [\zeta_{hk}]$ is the <u>incidence matrix from cells to branches</u>.

We have thus two incidence matrices: η from branches to nodes, and ζ from cells to branches. In our usual vector matrix notation (3.3) reads

$$\delta \underline{x}'\underline{e} = \underline{x}'\zeta\underline{b}. \tag{3.4}$$

In complete parallel to the branch case one defines as 2-cycles the chains with boundary zero, that is $\underline{x}'\underline{e}$ is a 2-cycle whenever $\underline{x}'\zeta\underline{b} = 0$ that is whenever

$$\underline{x}'\zeta = 0. \tag{3.5}$$

From Chapter II, Section 5 we have

$$\chi(K) = \alpha_0 - \alpha_1 + \alpha_2 = R_0 - R_1 + R_2 .$$

4. <u>Subdivision Invariance</u>

The importance of the homology groups H_i, $i = 0,1,2$ and of their dimensions R_i — Betti numbers — rests upon their being the same for two homeomorphic figures, for example a sphere S^2 and the boundary of a 3-cell say. This points to a more "important" relation between two figures. Actually it is that of <u>topological identity</u>. However, its full treatment calls for a deeper theory than we wish to utilize. We shall therefore confine our argument to "subdivision invariance", that is to the identity of various properties upon mere subdivision.

Now subdivision of a complex of dimension ≤ 2 is merely obtained by a repetition of simpler operations: <u>elementary subdivisions</u>. Hence we shall only require the treatment of invariance under elementary subdivision. For a graph G it merely consists of adding a node. For a 2-complex it may consist either in adding a new node or else just a new branch. I shall avoid a more extensive digression by suggesting the solution of this easy

<u>Problem</u>. <u>To prove that the homology groups</u> H_i, $i = 0,1,2$ <u>of a 2-complex</u> K <u>(hence also their Betti numbers) are invariant under elementary subdivisions, hence also under any subdivision</u>.

<u>General remark</u>. In the whole theory of chains and cycles the only properties of real numbers that have been utilized are those appropriate to rational numbers: addition, subtraction, multiplication, division. Therefore, one could as well consider coefficients from any field F. Actually we will have occasion to use, in Chapter VIII the <u>field mod</u> 2 made up of the symbols 0,1 under the rules: $0.0 = 0.1 = 1.0 = 1 + 1$; $0 + 1 = 1 + 0 = 1.1 = 1$. In this field $1 = -1$ and so one may dispense with negatives. There are obtained Betti numbers, written $R_i(2)$ or R_i mod 2, but no changes otherwise.

CHAPTER VIII

SURFACES

Up to the present we have not seriously limited the type of complex envisaged, that is the nature of the polyhedron Π. The class of particular interest for the application to networks, and indeed the most noteworthy class from any viewpoint is the class of surfaces.

1. Definition of Surfaces

Generally speaking a surface S is a connected figure with the property that each point of S has for neighborhood a 2-cell. This definition is however too general for our requirements, and so we shall restrict "surface" to the following:

(1.1) Definition. A surface is a connected polyhedron $|K|$ whose covering complex K has the following two properties:

A. Every branch b of K is adjacent to exactly two cells e, e' of K, and their boundary loops have only b in common.

Let n be any node of K and b_1 a branch ending at n. There is a cell e_1 adjacent to b_1 and so there is a second boundary branch b_2 of e_1 ending at n. Hence there is a second cell e_2 adjacent to b_2 and another branch b_3 of e_2 ending at n, etc. There results a sequence $b_1 e_1 b_2 e_2 \dots e_k b_k$, ending when b_1 occurs again, in which b_h is adjacent to e_{h-1} and e_h, $(e_{k+1} = e_1)$. This is a circular system of elements or umbrella with the common node n. Now *a priori* there may be several such systems attached to any node. For example, this happens if one takes

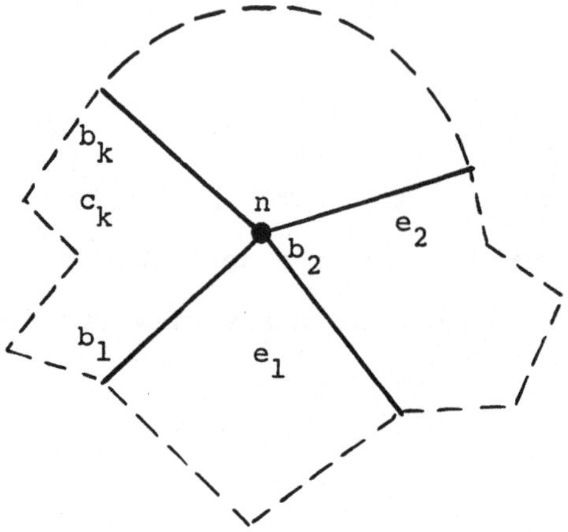

Figure 24.
Umbrella

the surface of a cube and brings two vertices into coincidence. However, we impose:

B. <u>To each node there is attached a unique umbrella</u>.

One may readily verify that each point of the surface has for neighborhood a 2-cell. In other words, our surfaces have adequate "smoothness".

(1.2) <u>Properties A,B are topological (proof omitted)</u>.

2. Orientable and Nonorientable Surfaces

The important property is really A. It implies immediately that

$$\gamma_0 = \Sigma e_h$$

is a cycle mod 2, since in the boundary $\delta\gamma_0$ each branch appears twice. There are now two possibilities: One may or one may not orient all the cells e_h so that γ_0 is actually a real cycle. In the first case the surface is orientable, in the second nonorientable.

Strictly speaking the orientability process refers to the covering complex K and not to the surface. However, the following theorem, stated without proof, justifies its direct assignment to the surface.

(2.1) Theorem. Orientability (hence also nonorientability) is a topological property.

In other words, the two properties are independent of the particular covering complex of the surface S.

Let the cell e_1 be oriented and let b_1 be a branch of the loop $\lambda(e_1)$, boundary of e_1. There is another cell e_2 adjacent to e_1, and let it be oriented so that e_1 and e_2 are oppositely related to b_1 that is so that b_1 does not appear in $\delta(e_1+e_2)$. Call this the "orientation process" (Figure 25). By this process if S

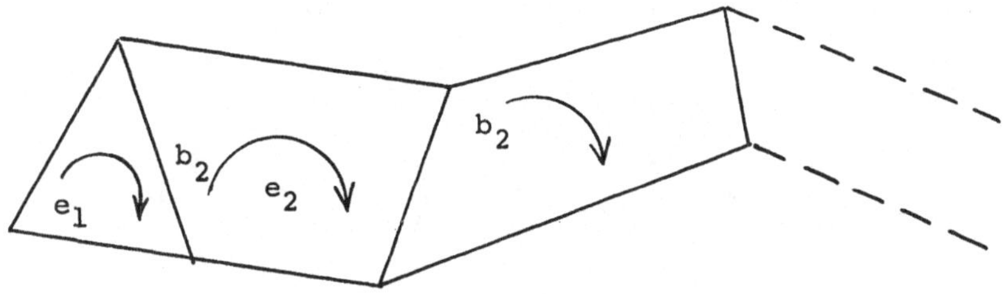

Figure 25.

is orientable one will obtain a unique orientation for all the cells.

That is any geometric chain such as Figure 26 will behave as indicated: there will be no violation of the orientation process. On

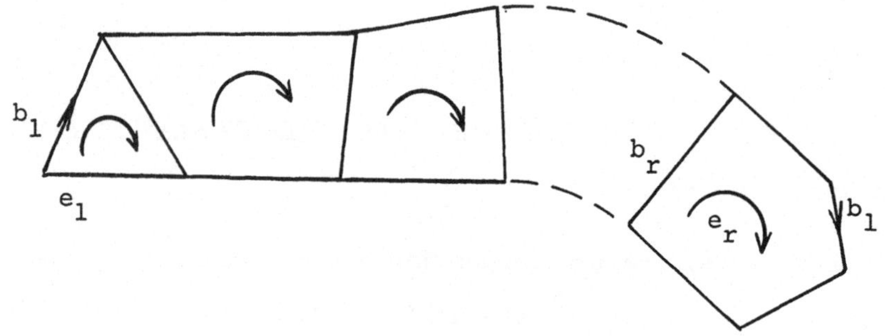

Figure 26.

Orientable Geometric Chain

the contrary in a nonorientable surface there will always be some geometric chain of the type of Figure 27.

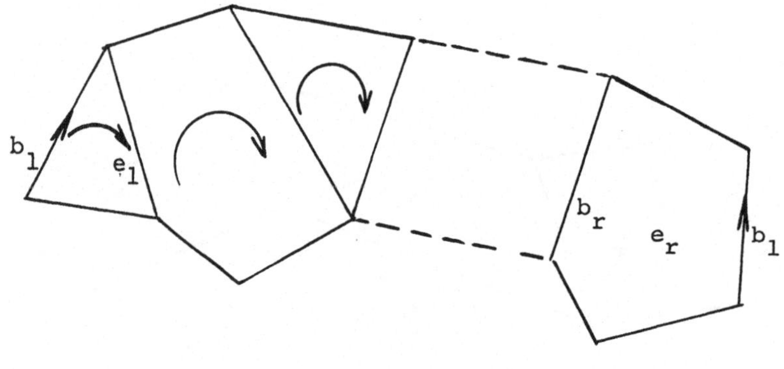

Figure 27.

Nonorientable Geometric Chain

In the orientable case all geometric chains drawn on the surface behave like a cuff. In the nonorientable case there will be some band of the type of Figure 28: the sides 1,2 are matched as indicated:

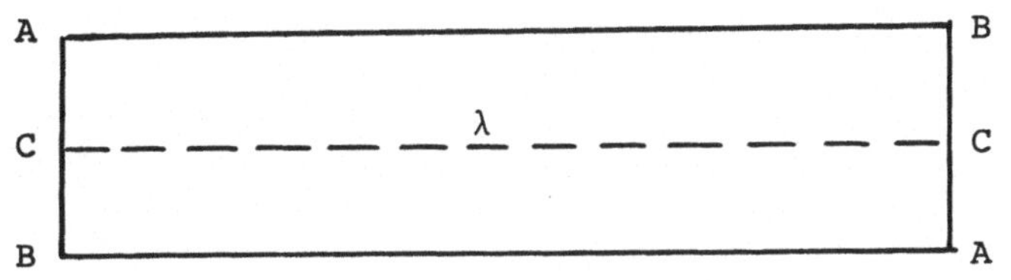

Figure 28.

Möbius Strip

This is the classical Möbius strip, the first example (given by Möbius over a century ago) of a nonorientable surface piece. (See also Chapter III, Section 1.)

A noteworthy property is this:

(2.2) Theorem. All the 2-cycles of an orientable surface are of the form $x\gamma_0$. The only 2-cycle mod 2 on a nonorientable surface other than zero is γ_0 itself.

Suppose that S is orientable and let

$$\gamma_1 = x_1 e_1 + x_2 e_2 + \cdots + x_{\alpha_2} e_{\alpha_2}$$

be any 2-cycle. Suppose that $x_1 \neq 0$. Then between e_1 and e_r one may draw a geometric chain such as in Figure 26. Since γ_1 is a cycle $\delta\gamma_1$ will not contain b_1. Since the coefficient of b_1 in $\delta\gamma_1$ is $\pm(x_1 - x_2)$ we must have $x_2 = x_1$, and likewise $x_2 = x_3 = \ldots = x_r$. Hence $\gamma_1 = x_1 \gamma_0$.

The same proof holds for the nonorientable case, except that every $x_h = 1$.

Betti numbers. A noteworthy consequence of the above is this distinction between orientable and nonorientable surfaces:

$$\left.\begin{array}{l}\underline{\text{Orientable}}\\ \underline{\text{Nonorientable}}\end{array}\right\} \underline{\text{surface}} \quad R_2 = \begin{cases}1\\0\end{cases}.$$

That is, one may distinguish between the two by the value of R_2.

Suppose that K is nonorientable. Take for each closed cell e two copies e', e'' oriented in opposite ways. Let e_h, e_k be adjacent across the branch b. Unite say e'_h with the one of the cells $e'_k e''_k$ which is oppositely related to b. As a consequence, and since K is nonorientable there results a new complex K^* which is a surface (proof elementary) and is orientable. The surface K^* is said to be a doubly covering surface of K.

Note that the same procedure applied to an orientable surface K results in two distinct orientable surfaces K_1^* and K_2^* with corresponding cells oppositely oriented and both copies of K.

Returning again to the nonorientable surface K, since the numbers of nodes, branches, cells in K^* are double the same for K:

$$\chi(K) = \frac{1}{2}\chi(K^*).$$

3. Cuts

Let λ be a loop in the surface S. The umbrella construction of Section 1 may be applied to the cells with a node or branch in λ, with λ replacing the node n in the construction. The resulting number q of umbrellas cannot exceed two. For if $q > 2$ some branch of λ would be adjacent to q cells, and S would not be a surface. Thus $q = 1$ or 2.

Suppose that $q = 1$: there is just one umbrella. Upon applying the umbrella construction there is an associated description of the loop λ in a certain direction. As one starts from a node n in that direction and with cell e, when one returns to n and passes it one will arrive a cell $e' \neq e$, for $e' = e$ would imply that there is an umbrella which does not include all the cells with node or branch in λ, and hence that there are two umbrellas. As a consequence the orientation scheme breaks down as between e and e'. Hence the surface S is nonorientable.

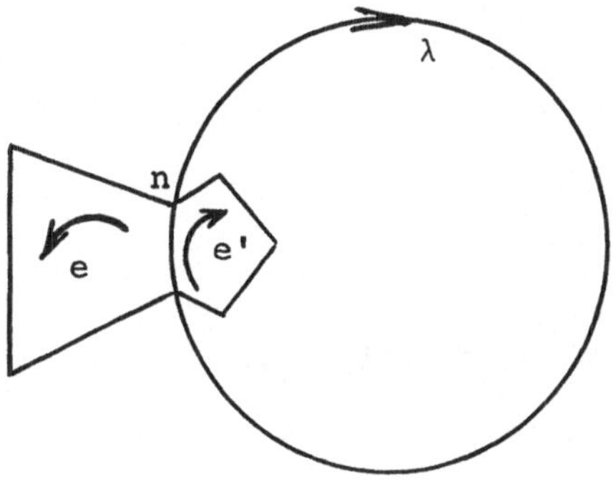

Figure 29.

One-sided Cut

Conversely, let S be nonorientable. The full orientation process breaks down along a collection of branches making up a graph H. A node n of H cannot be of order one, since manifestly the orientation process cannot break down around it. It follows that H contains no tree. Hence H contains a loop λ around which the orientation scheme breaks down. However, this can only happen if λ has a single umbrella. Thus

(3.1) Theorem. A n.a.s.c. in order that a surface be non-orientable is that it possess a loop with a single umbrella.

(3.2) Corollary. A n.a.s.c. in order that a surface be orientable is that every loop possess two umbrellas.

A loop λ is said to be a 2-sided [one-sided] cut whenever it possesses two umbrellas [one umbrella]. The umbrellas are the sides of the cut.

The remainder of our argument will rest heavily upon a set $\lambda_1,\dots,\lambda_s$ of disjoint loops which do not disconnect the surface S. In this connection we prove the following basic property:

(3.3) Let p be a maximum number of disjoint nondisconnecting cuts. Then for an orientable [nonorientable] surface $p \le R_1$ [$p \le R_1 \bmod 2$].

Let first S be orientable and let $\lambda_1,\lambda_2,\dots,\lambda_s$ be a set of oriented disjoint nondisconnecting cuts of S. Designate also their cycles by λ_h.

(3.4) The cycles λ_h are independent.

Assume a boundary relation

$$\delta \Sigma \beta_h e_h = \Sigma \gamma_k \lambda_k . \tag{3.5}$$

Since $K - \Sigma \lambda_h$ is connected the reasoning of Section 2 will apply here and we will find that with suitable orientation of the cells e_h: $\beta_1 = \beta_2 = \dots = \beta$. Hence

$$\delta \Sigma \beta_h e_h = \beta\delta \Sigma e_h = \beta\delta\gamma_0 = 0 = \Sigma \gamma_h \lambda_h ,$$

where γ_0 is the same as in Section 2. Since no two λ_h have common branches we must have every $\gamma_h = 0$, that is (3.5) is trivial.

Since the λ_h are independent cycles $s \leq R_1$.

When S is nonorientable the reasoning is the same save that chains and cycles are to be taken mod 2 and so $s \leq R_1$ (mod 2).

Let K_1 be an elementary subdivision of K and $\lambda_1, \ldots, \lambda_s$ as above. It is immediately seen that these loops give rise to a similar set for K_1. Hence the maximum number s(K) of such loops is $\leq s(K_1)$, the same number for K_1. Hence:

(3.6) Subdivision does not decrease s(K).

Since subdivision does not affect the Betti numbers R_1 or R_1 mod 2, there is a subdivision, call it again K_1, for which $s(K_1) = p \leq R_1$ or R_1 mod 2, where it is the largest possible value of $s(K_1)$. In the future K will be replaced by K_1. This amounts to assuming, as we shall do henceforth that the maximum p has already been attained by K itself.

4. A Property of the Sphere

The object of the present section is to prove this basic result.

(4.1) Theorem. If every cut disconnects a surface S then S is a sphere.

Strictly speaking this theorem is implicit in the Jordan-Schoenflies theorem (Chapter II, Section 7). However, we merely aim to give a relatively elementary proof of our special case.

Since onesided cuts do not disconnect the assumption implies:

(4.2) S is orientable.

The proof of (4.1) rests upon an induction on the number of

cells of the complex K. Let P_m represent (4.1) for a surface with a number m of cells. To prove P_m we need to prove the following properties:

(4.3) _If_ m = 2, K _is a sphere_.

(4.4) P_{m-1} _implies_ P_m.

Proof of (4.3). When m = 2 one may represent K as an ordinary Euclidean sphere whose cells are the upper and lower (open) hemispheres and common boundary the equator E divided into $q \geq 3$ arcs by 3 nodes. Thus K is a sphere.

Proof of (4.4). Let K have m > 2 cells. Take one of these e and replace it by a point A. This is done as follows. If λ is the boundary loop of e, it has an umbrella U not containing e. Let K_1 be the complex consisting of A and the cells of K - e plus their boundaries modified as follows: if e_1 is a cell of U with a node or branch in λ then they are replaced by A.

In K_1 the umbrella U becomes the unique umbrella of A. Since each branch of K_1 is still adjacent to just two cells, and K_1 is manifestly connected, it is a surface and has exactly m - 1 cells. If we can show that every loop of K_1 disconnects K_1 then by P_{m-1} it will be a sphere. Hence $P_{m-1} - A$ is a cell (proof immediate) and so is $P_m - e - \lambda$. Hence by (4.3) P_m is a sphere. Thus our final task is to prove:

(4.5) _In_ K_1 _every cut_ μ _disconnects_.

Before proceeding with the proof of (4.5) observe that $V = e + \lambda + U$ is a cell. For it has exactly the structure of a convex closed plane polygon plus an umbrella around it. Hence, in particular, any two points of V may be joined by an arc in V.

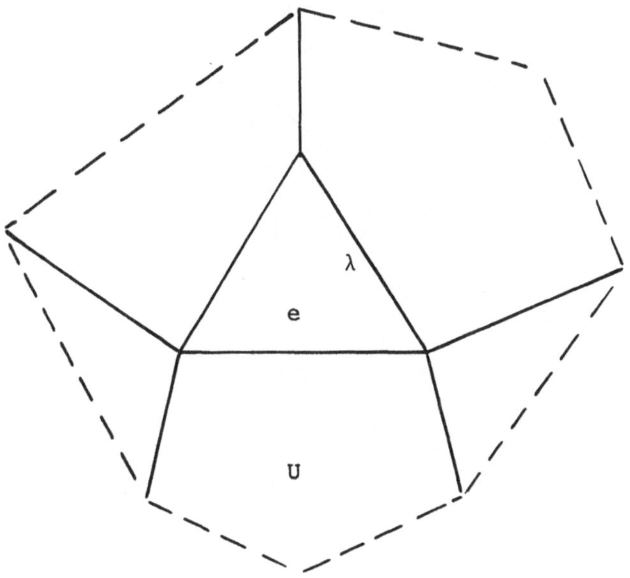

Figure 30.

Proof of (4.5). Suppose first that $K_1 - A - U$ is a point B. Then, manifestly K_1 is a sphere. In that case again $K_1 - A$ is a cell. Hence K consists of two cells with common boundary λ and so K is a sphere. Hereafter then we may suppose that $K_1 - A - U$ is not a point. Under these conditions it will be shown that the assumption that a cut μ of K_1 does not disconnect K_1 leads to a contradiction.

Suppose first that μ passes through A (Figure 31). Then contains two branches $b = n'A$, $b' = n''A$ where n', n'' are nodes opposite A in U. Since U is an umbrella one may join n' to n''

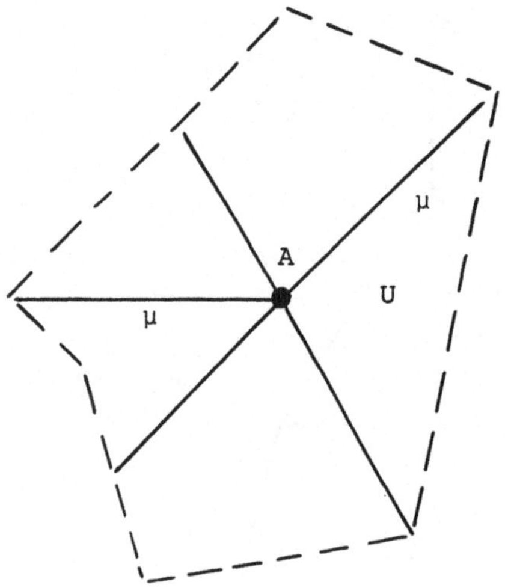

Figure 31.

by a polygonal line ν in the boundary B of U. If μ' is the result of replacing n'An" by ν in μ then μ' does not disconnect. Therefore, it contains - a loop - a cut - μ'' which does not disconnect and does not contain A. Therefore, we may assume that the initial cut μ does not contain A. Under the assumptions μ has no points in U + A. Any pair of points x,y of $K_1 - \mu$ may be joined by an arc ν. One may manifestly assume that neither x nor y is the point A, and in fact that ν does not contain A. Hence μ is the image of a cut $\overline{\mu}$ in K and x,y a pair imaged into $\overline{x},\overline{y}$ of $K - \overline{\mu}$ joined by an arc in $K - \overline{\mu}$. Since $\overline{x},\overline{y}$ are merely arbitrary points of $K - \lambda - e - \mu$ it is readily shown that this contradicts the assumption P_m. Hence (4.5) is proved and so is the theorem.

5. Reduction of Orientable Surfaces to a Normal Form

Let K be an orientable surface. Assume that the maximum possible p of disjoint nondisconnecting cuts for K and its subdivisions is already reached for K. If it is not one replaces K by a subdivision for which p is reached and call K this subdivision. Let $\lambda_1, \lambda_2, \ldots, \lambda_p$ be a set of disjoint nondisconnecting cuts of K. Take any cut λ_h and let U_{h1}, U_{h2} be its two umbrellas. Construct a new complex K^* in which the two umbrellas are untouched but each λ_h is replaced by an identical reporduction of λ_h, say λ_{h1} and λ_{h2} for U_{h1} and U_{h2}. Construct new cells e_{h1} and e_{h2} with respective boundaries $\lambda_{h1}, \lambda_{h2}$. The complex K^* results from this operation for each h. One verifies immediately that K^* is an orientable surface.

(5.1) K^* _is a sphere_.

The proof is the same, with minor modifications of that of (4.1), and need not be repeated.

It follows that

$$\chi(K) = \chi(K^*) - 2p = 2 - 2p. \tag{5.2}$$

Hence if K is not a sphere $\chi(K) < 2$. Since if K is nonorientable $\chi(K)$ is half the characteristic of a certain orientable surface its $\chi < 2$ also. Thus:

(5.2) _Theorem_. _If_ K _is not a sphere its characteristic_ $\chi(K) < 2$. _Thus the sphere is completely characterized by the value 2 of its characteristic_.

The number p is known as the _genus_ of the surface S.

The normal form S_p corresponding to p is obtained by cutting out p pairs of holes $\lambda_1, \lambda_1', \ldots, \lambda_p, \lambda_p'$ in a sphere and covering each pair with a cylindrical "handle". If one admits, as we shall, that p is a topological invariant, the models obtained for distinct values of p are all distinct.

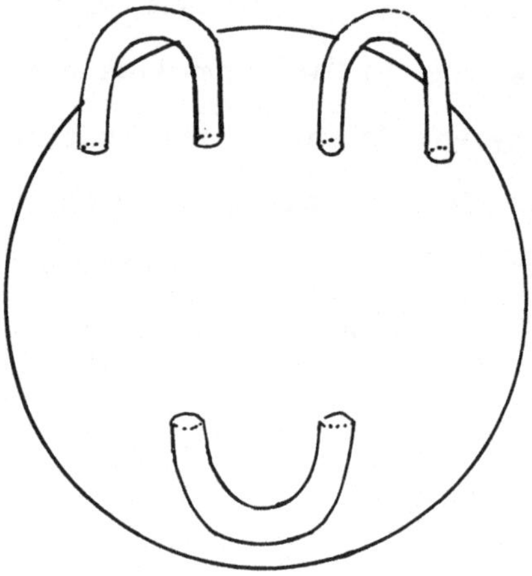

Figure 32.

Sphere with 3 handles

6. Reduction of Nonorientable Surfaces to a Normal Form

The simplest type of nonorientable surface is the projective plane P. Topologically it is obtained from a circular region with identification of diametrally opposed points (Figure 33). Its triangulation is indicated in Figure 33. The count of elements gives the values $\alpha_0 = 6$, $\alpha_1 = 15$, $\alpha_2 = 10$. Hence $\chi(P) = 1$. One verifies readily that P is a surface and since its characteristic is odd it is nonorientable. This can also be ascertained from the presence in Figure 33 of a Möbius strip (shaded region). The covering surface

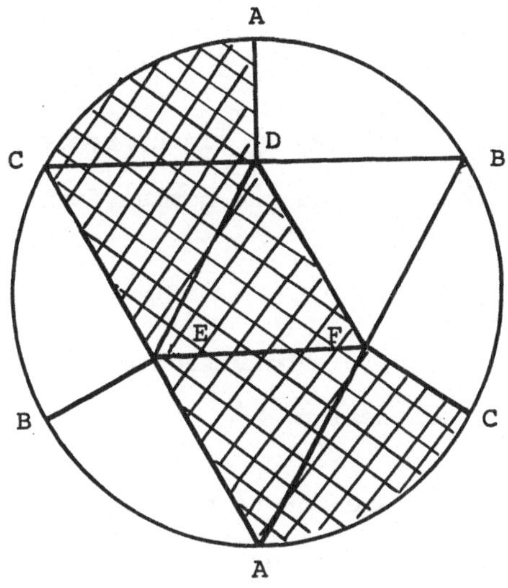

Figure 33.

Projective Plane

has characteristic 2 and so it is a sphere. Hence P is the nonorientable surface with largest characteristic.

We have seen that a nonorientable surface S must have a Mobius strip, through a one-sided cut. Let it be λ_1 and let $\lambda_1, \lambda_2, \ldots, \lambda_q$ be a maximal set of disjoint nondisconnecting cuts. As before we may assume that q is the largest possible for all the subdivisions of S.

Suppose now that λ_2 is a two-sided cut. We replace it by a cut λ_2' such as outlined in Figure 33, and λ_2' will be one-sided. The set $\lambda_1, \lambda_2', \lambda_3, \ldots, \lambda_q$ continues to be nondisconnecting. By this process the initial set will be replaced by a set still called

$\lambda_1, \lambda_2, \ldots, \lambda_q$ consisting only of one-sided cuts. Let this already be achieved.

If one cuts open the loops and covers them with cells, one will have a sphere as before. Hence

$$X(S) = X(S_q) = 2 - q.$$

The return to the initial surface S_q is by replacing the cells by projective planes and this will be the normal form.

7. Duality in Surfaces

Let K' be a barycentric subdivision of the surface K. Introduce the following collection K^* of elements (Figure 34):

I. Nodes n_k^* in one-to-one correspondence with the cells e_k of K; n_k^* is the centroid of e_k.

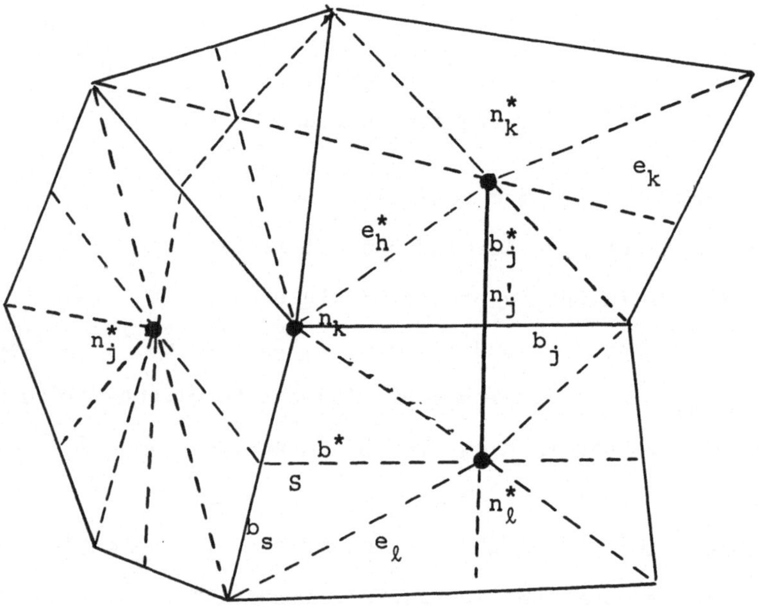

Figure 34.

II. Branches b_j^* defined as follows. The branch b_j is adjacent to two cells e_k, e_ℓ of K. On b_j there is a new node n_j' of K' and on e_h, e_ℓ new nodes n_k^*, n_ℓ^* of K^*. The branch $b_j^* = n_k^* n_j' + n_j' + n_j' n_\ell^*$.

III. Cell e_h^* consisting of the node n_h plus all the open branches and triangles of K' ending at n_h.

It is a simple matter to verify the following properties.

(a) The collection K^* is a surface.

(b) K' is a barycentric-like subdivision of K^* the new vertices are not necessarily "centroids", but the structure of K' relative to K^* is that of a barycentric subdivision.

It follows from (b) that

$$R_1(K^*) = R_1(K') = R_1(K). \tag{7.1}$$

<u>Hence if</u> K <u>is orientable [nonorientable] so is</u> K^*.

Suppose that K <u>is orientable</u>. Then K' is likewise orientable. Hence the orientation of any triangle of K' is determined by that of any one triangle plus the orientation process. One may then determine through the orientation process the orientations of all the cells of K, K', K^*. Moreover, as a consequence the analogues of the matrices η, ζ for K^* are merely ζ', η'. Hence the one-one relation $n_j, b_h, e_\ell \longleftrightarrow e_j^*, b_h^*, n_\ell^*$ determines a duality relation $K \longleftrightarrow K^*$ relative to the operations δ, δ^* in the sense of II.

CHAPTER IX

PLANAR GRAPHS

1. Preliminaries

A graph G is planar if one may sketch it on an Euclidean plane, that is represent it faithfully (map it topologically) on the plane: The problem of finding conditions under which G is planar is of evident geometric interest, and also of interest for network theory as we shall see. This is the main problem to be discussed in the present chapter.

The first solution was given by Kuratowski who not only solved the graph problem but found n.a.s.c. for a continuous curve Γ: map of an interval, to be planar. A more strictly "graph" solution was first given by Hassler Whitney and was improved a few years later by Saunders MacLane.

Now both Whitney's and MacLane's solutions were strictly "graphic" and of the type known as combinatorial - that is with topological considerations dispensed with. I have shown that by a judicious use of the theory of surfaces as developed in Chapter VIII, one could obtain a far more rapid solution. This is the solution that will be given here. Within the same order of ideas it has been found possible to deal with a noteworthy extension of the graph problem.

From many points of view the Euclidean plane is a rather awkward and sometimes exasperating figure. This is due mainly to its infinite extension. Now if a graph G may be sketched on a plane it may likewise be sketched on a sphere and conversely. My deviation from my predecessors is mainly due to having dealt with the graph-sphere problem: to find n.a.s.c. in order that a graph may be sketched upon a sphere.

Relevant Bibliography

Kuratowski, Casimir

"Sur le problème des courbes gauches en topologie" Fundamenta Mathematicae, Vol. 15, 1930, pp. 271-283.

Whitney, Hassler

"A set of topological invariants for graphs" American Journal of Mathematics, Vol. 55, 1933, pp. 231-235.

"On the classification of graphs" American Journal of Mathematics, Vol. 55, 1933, pp. 236-244.

"Planar graphs" Fundamenta Mathematicae, Vol. 21, 1933, pp. 73-84.

MacLane, Saunders

"A structural characterization of planar combinatorial graphs" Duke Mathematical Journal, Vol. 3, 1937, pp. 460-472.

Lefschetz, Solomon

"Planar graphs and related topics" Proceedings of the National Academy of Sciences, Vol. 54, 1965, pp. 1763-1765.

2. Statement and Solution of the Spherical Graph Problem

Let G be our usual graph with its nodes n_h and branches b_k. Suppose that G has a topological image G_1 on a Euclidean sphere S. It is an easy matter to show that $S - G_1 \neq 0$. Take then a point A of $S - G_1$ and from A project G_1 on a plane Π not passing through A. The projection G_2 is a topological image of G in Π. Conversely given a topological image G_2 of G in Π and a point A of S not in the intersection $S \cdot \Pi$ the projection G_1 of G_2 from A onto S is a topological image of G_1 in S. Thus to find conditions under which G is planar or spherical are equivalent problems.

We shall deal directly with the spherical problem.

A natural restriction on G is to assume that it is connected.

A less evident restriction is the property of separability introduced by Whitney. A connected graph is separable whenever the removal of some node (not any node) disconnects it. It is not difficult to show that if G is disconnected by the removal of a node into G' and G" and if the closures $\overline{G}'$ and $\overline{G}''$ are spherical graphs so is G. Therefore we simplify matters by assuming definitely that the graph under consideration is inseparable that is both connected and inseparable.

The best planar graph theorem is due to Saunders MacLane, and this is the proposition that we propose to prove. However, because of our looking at it as a spherical theorem our phrasing differs somewhat from MacLane's original version.

(2.1) Theorem of Saunders MacLane. *Let G be connected and inseparable with Betti number R. N.a.s.c. in order that G may be represented as a spherical graph is that it possess a set of $R + 1$ loops $\lambda_1, \lambda_2, \ldots, \lambda_{R+1}$ such that*

I. Every branch of G belongs to exactly two loops λ_h.

II. Let λ_h also designate the cycle of λ_h. Then with a suitable orientation of the cycles λ_h the only independent relation which they satisfy is

$$\Sigma \lambda_h = 0. \tag{2.2}$$

Property II implies that any R of the λ_h form a base for the cycles of G.

Proof of necessity. In order not to interrupt the main argument we will prove first this elementary property:

(2.3) <u>A graph</u> G <u>has only a finite number of geometrically distinct loops</u>.

For the loops are in one-one correspondence with some of the symbols $(b_h, b_j, \ldots, b_k)$ whose number is manifestly finite.

Suppose now that G connected and inseparable is spherical and let it be identified with its image in an Euclidean sphere S. On the strength of the Jordan Schoenflies theory (Chapter III, Section 7) plus the connectedness and inseparability of G we find readily enough that $S - G$ consists of cells bounded by loops of G. Hence the number $1 + R'$ of the cells is finite. Designate them by $e_1, e_2, \ldots, e_{1+R'}$ and let $\lambda_h = \delta e_h$. Thus $G + \Sigma\, e_h = K$, a covering complex K of the sphere S.

Recall now the following topological invariances:

(a) of properties A,B of the definition of surface (Chapter VIII, Section 1);

(b) of the Betti numbers (Chapter VII, Section 4).

As a consequence K defines an orientable surface and $R_2(K) = 1$. Hence if the cells e_h are properly oriented

$$\gamma_0 = \Sigma\, e_h \tag{2.4}$$

is a 2-cycle of K. Moreover, any other 2-cycle $\gamma = \alpha\gamma_0$. Hence first

$$\delta\gamma_0 = \Sigma\, \lambda_h = 0, \quad 1 \leq h \leq 1 + R'. \tag{2.5}$$

Suppose that there is a relation

$$\Sigma\, \alpha_h \lambda_h = 0. \tag{2.6}$$

This implies that

$$\delta\, \Sigma\, \alpha_h e_h = 0,$$

or that $\Sigma\, \alpha_h e_h$ is a 2-cycle. Hence

$$\Sigma\, \alpha_h e_h = \alpha\, \Sigma\, e_h,\ \alpha_1 = \alpha_2 = \dots = \alpha.$$

Hence (1.5) is a consequence of (1.4). Thus except for the value of R' the λ_h already possess property II. Let, however, λ be any loop of $G = S$ since $R_1(K) = 0$ necessarily

$$\lambda = \delta\, \Sigma\, \beta_h e_h = \Sigma\, \beta_h \lambda_h,$$

hence R' of the λ_h form a base for all the 1-cycles of G. Therefore $R' = R$. Thus the λ_h have property II.

Regarding property I let

$$H = \Sigma[\lambda_h],$$

so that H consists of all closed branches in the λ_h. Assume that $L = G - H \neq 0$. Then some branch b_1 of L must have a node n_1 in H, for otherwise L would be a subgraph of G disjoint from H, in contradiction to the connectedness of G. Let n_2 be the other node of b_1 and suppose that it is in L. Then there is a second branch b_2 in L with a node $n_3 \neq n_2$, etc. Thus there arises a sequence $n_1\, b_1,\, n_2\, b_2 \dots b_{r-1}\, n_r$ with n_1 and perhaps also n_r in H and the rest of the terms in L. The process must end and can only do so in one of these three ways:

(a) n_r is in H. This gives rise to a new loop μ <u>not</u> in H. This is excluded since H by construction contains all the loops of G.

(b) $n_s = n_r$, $s < r$. Same objection.

(c) n_r is of order unity. This is excluded since G is inseparable: it would break down through the removal of the node n_{r-1}, the other node of b_{r-1}.

The conclusion is that $L = 0$, $G = H$, that is every branch b

of G is part of a loop of G. Because K defines a surface b is part of exactly two loops of G. Thus property I is also satisfied. This completes the proof of necessity.

Proof of sufficiency. Let $\{\lambda_k\}$ be the collection of loops of G fulfilling properties I and II of the theorem. Cover the λ_k with cells e_k so oriented that $\delta e_k = \lambda_k$ and let K be the resulting complex. K is clearly connected and has property A of Chapter VIII, Section 1. Hence what may prevent K from being a surface is the possible nonuniqueness of umbrellas at some of the nodes. Suppose that there are p umbrellas $(p > 1)$ $U_1, U_2, \ldots, U_p$ at the node n. Construct a new complex K' as follows. Outside of the U_j it coincides with K. The umbrella U_j is reconstructed as before save that its central node n is replaced by a new node n_j'. This construction is repeated in succession for all the nodes with multiple umbrellas. There results a new complex K^* which is manifestly a surface. However, if ε is the number of excessive umbrellas in K (for the node n above it was $p - 1$), then K^* has merely ε more nodes than K. Hence

$$\chi(K) = \chi(K^*) - \varepsilon < 2,$$

if $\varepsilon \neq 0$.

On the other hand since the number α_2 of cells of K is $R + 1$ we have

$$\chi(K) = \chi(G_R) + R + 1 = 1 - R + R + 1 = 2.$$

Hence $\varepsilon = 0$ and K is a surface. Since $\chi(K) = 2$, K covers a sphere S. Hence G is spherical. This proves sufficiency and hence also the theorem.

3. Generalization

This generalization refers to the possible imaging of a given graph G into a surface other than a sphere. However, since one will cease to dispose of the simple property $\chi(S) = 2$, more strict conditions will have to be imposed.

If K is any complex (not necessarily a surface) the graph obtained by removing the cells of K is called the skeleton of K. We shall only aim at the outset to image G as the skeleton of a covering complex of a surface. Here one must distinguish between orientable and nonorientable surfaces.

Take first an orientable surface S_p, of genus $p > 0$. Let $\lambda_1, \ldots, \lambda_s$ be a set of loops of G. As in Chapter VII, form a complex $K(\lambda)$ by introducing a set of disjoint cells e_h, where $\delta e_h = \lambda_h$. We may then state:

(3.1) Theorem. In order that a connected inseparable graph G may be imaged as a skeleton of a covering complex K of a surface S_p n.a.s.c. are that it contain a set of $R - 2p + 1$ loops $\lambda_1, \lambda_2, \ldots, \lambda_{R-2p+1}$ such that:

I. Every branch of G belongs to exactly two loops λ_k.

II. The loops with any single common node n_k form a "graph-umbrella": system $b_1\lambda_1, b_2\lambda_2, \ldots, b_r$, $(b_r = b_1)$ where b_h is adjacent to λ_{h-1} and λ_{h+1}.

III. The only independent relation satisfied by the cycles λ_h is

$$\Sigma \lambda_h = 0. \tag{3.2}$$

Proof of necessity. Identify G with its image as skeleton of a covering complex K. As for the sphere this complex defines an orientable surface and $R_2(K) = 1$. If e_h are its cells properly oriented then

$$\gamma = \Sigma e_h$$

is the only independent 2-cycle of K. If $\lambda_h = \delta e_h$, the λ_h are loops of G and one shows as in Section 1 that I and III hold, while II is a consequence of the fact that K covers a surface.

Proof of sufficiency. If I, II, III hold for G then $K(\lambda)$ is a surface. Then

$$\begin{aligned} \chi(K(\lambda)) &= \chi(G) + R + 1 - 2p \\ &= 1 - R + R + 1 - 2p = 2 - 2p. \end{aligned}$$

Hence $K(\lambda)$ is an S_p in which G is imaged as affirmed.

Consider now a nonorientable surface S_q with characteristic $1 - q$. Since $R_2(S_q) = 0$, $R_0(S_q) = 1$, we have $(1-q) = 1 - R_1(S_q)$, hence $R_1(S_q) = q$. Thus q is the first Betti number of the surface. The related result is:

(3.3) Theorem. Same statement as (2.1) with these modifications: nonorientable S_q is now the surface; the loops are $\lambda_1, \lambda_2, \ldots, \lambda_{R-q}$ and they are independent.

The treatment is the same as in the orientable case and need not be repeated.

4. Direct Characterization of Planar Graphs by Kuratowski

As already stated the characterization of Kuratowski, the earliest in date, is also applicable to more general figures than graphs.

(4.1) Theorem of Kuratowski. N.a.s.c. in order that a graph be planar is that it fail to contain the topological image of one of the following two graphs:

(a) graph made up of the edges of a tetrahedron plus a segment joining two opposite edges (Figure 35);

(b) graph consisting of 5 nodes joined in pairs by branches in all possible ways (10 branches) (Figure 36).

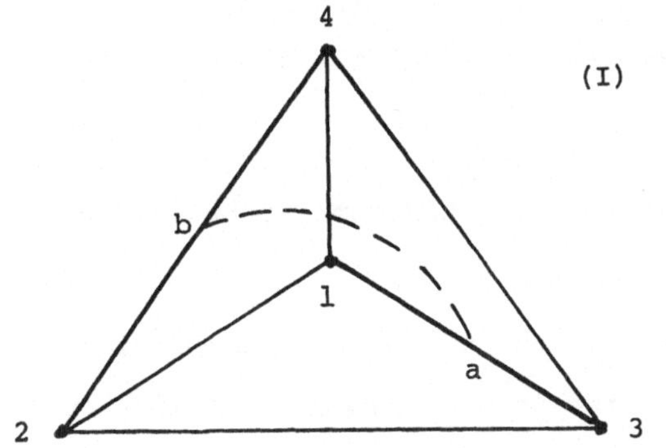

Figure 35.

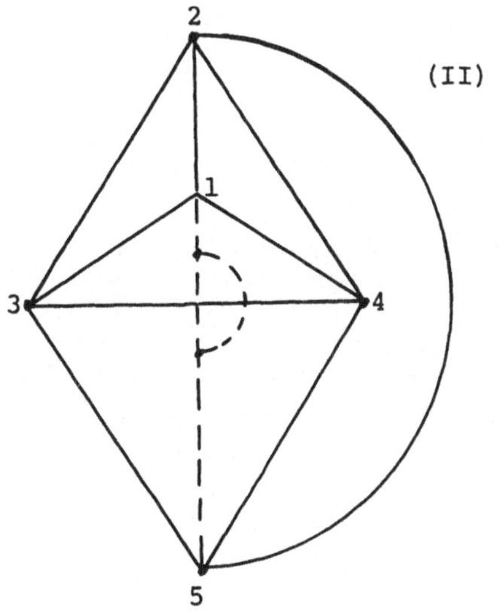

Figure 36.

We shall prove directly the following equivalent formulation of:

(4.2) N.a.s.c. for a graph to be planar is that it does not contain topological images of the types I or II.

We shall actually obtain certain intermediary forms which arise as variants of I or II. We examine them before taking up the proof of the theorem.

Variants of I. They could come up by interchanging in 5 vertices 1 and 2, or 3 and 4 or making $1 = 2$, or $3 = 4$, or both. These variants are readily drawn and verified to be planar. As we shall see they will not arise in our argument. A very curious type arises for $1 = 2,\ 3 = 4$ and may be represented in Figure 37 by the intersection of two circles plus a dotted arc. However, the equivalent graph of Figure 38 is planar.

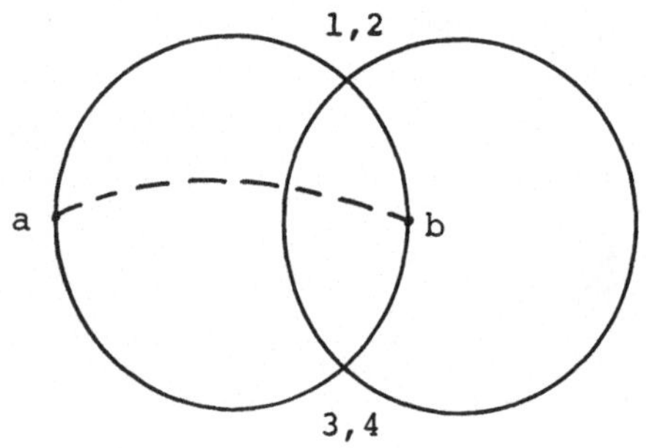

Figure 37.

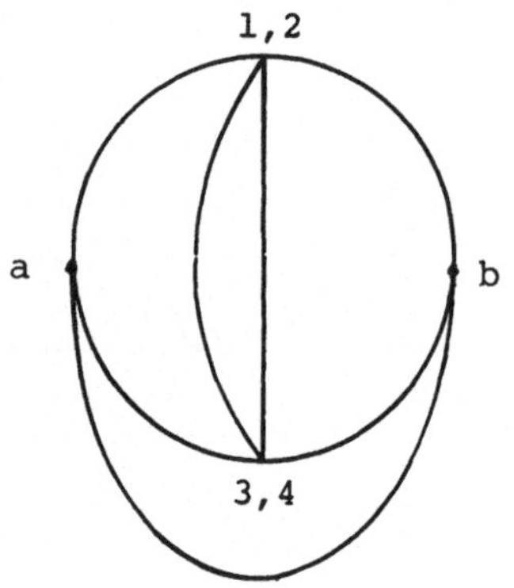

Figure 38.

Variants of II. These variants are obtained by substituting for the arc ρ an arc with endpoints near 1 and 2. All these variants contain I as graphs and they may well arise later. Let us show that they do effectively contain I as graph. The designations will refer to Figure 39. Distinguish essentially two cases: (a) 1' and 2' are both to the right of 1 and 2, as shown in Figure 39.

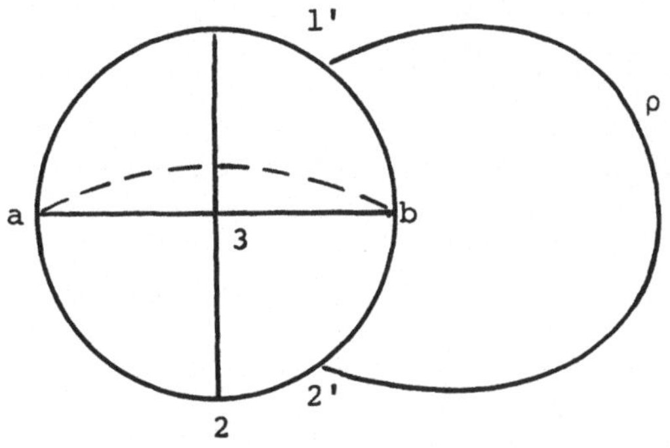

Figure 39.

The identification is then clear. (b) Say 1' is as before or else coincides with 1 and 2' is to the left of 2. The identification with I is again plain.

To sum up then:

(4.3) The variants which are not planar all contain I as graph.

(4.4) Proof of Kuratowski's Theorem. The method to be followed is really the same as Kuratowski's as adapted to graphs.

Two preliminary lemmas are required.

(4.5) Lemma. A tree T is always planar.

As usual let α_0 and α_1 be the number of nodes and branches of T. For $\alpha_1 = 1$, hence $\alpha_0 = 2$ the result is obvious. Given $\alpha_1 > 1$ let the assertion hold for a tree T' with $\alpha_1 - 1$ branches. Remove from T a branch b ending at a node n of order one together with the node n. The new graph T' is still a tree, hence planar since it has only $\alpha_1 - 1$ branches. One may then obviously add b as an arc in the complement of T' in the plane, likewise its endpoint n. The result is a representation of T as a planar graph.

(4.6) Lemma. A nonplanar graph G contains a pair of loops λ, λ' with just one common closed arc.

The resulting figure is what Ayres has called a θ graph, that is one made up of 3 arcs with common endpoints. Lemma (4.6) is in fact a special case of a general proposition due to Ayres, in which the graph G is replaced by a continuous curve.

(a) If two loops λ, λ' have more than one point in common then the graph contains a θ subgraph.

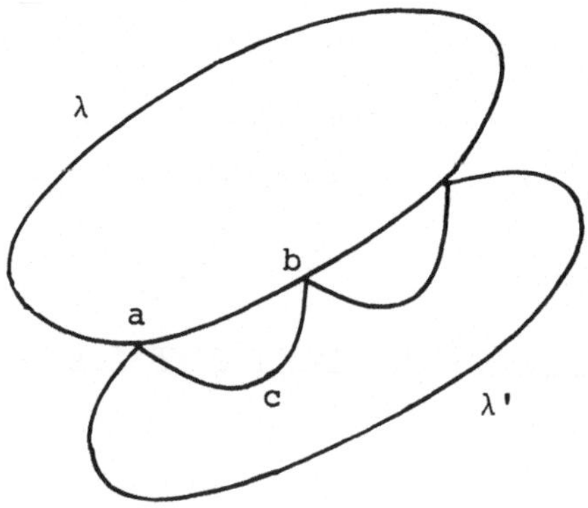

Figure 40.

Figure 40 represents two loops λ, λ' with more than one common point. Evidently λ plus the loop a c b make up a figure θ.

(b) If any two loops of a graph G have at most one point in common then G is planar.

We may evidently assume G connected.

Suppose first that G has a loop-cluster, that is a set of loops $\lambda_1, \ldots, \lambda_s$ with the following property:

(a) there is a point P such that either λ_h has the node P or else is connected with it by an arc;

(b) the loop-cluster meets no other loop of G. Let G' be the graph G - cluster + P.

If we can show that G' is planar, say in a plane Π, then by inserting the cluster in the obvious way in Π there will follow a planar representation of G itself. We may, therefore, assume that G contains no loop-cluster. As a consequence any loop λ of G has at least two nodes a,b each belonging to some other loop or attached to it by an arc. Let all the loops be oriented in some manner. Take some loop λ_1 and a node a of λ_1 which is also a node of a loop λ_2 or is attached to such a loop by an arc a_1a_2. In the first case we may consider that $a_1 = a_2$. Follow λ_2 positively until a new node a_3 is reached to which there is attached a loop λ_3 either directly or by means of an arc, etc. Suppose that the process either forward or backward from λ_1 never leads back to some λ_j, $j \geq 1$. Since the number of loops is finite we thus obtain a finite loop string. Let it consist of k loops and call it S_k. If $k = 1$ the string is planar. Assume that S_{k-1} has been shown to be planar. Then a kth loop may be attached in the obvious way to S_{k-1} in its plane and produce S_k as planar. Thus every S_k is planar.

Suppose now that we do not have a string. Then the kth loop, say forward is again λ_j, $j < k$ and we may as well take it to be λ_1, and $k \geq 3$. The successive arcs a_1a_2, $a_2a_3,\ldots$, give rise to a new loop μ. If the kth step leads to the same initial point a_1 as before then μ has an arc in common with λ_2. If it leads to a node $a_1' \neq a_1$ on λ_1 then λ_1 and μ have the common arc $a_1'a_1$. Since the case under present consideration corresponds to the only nonplanar possibility Lemma (3.5) is proved.

Proof of Kuratowski's Theorem (4.1). Let G be a nonplanar graph. Suppose that G has a nonplanar subgraph G_1 whose Betti number $R(G_1) < R(G)$. Let G_1 itself have a nonplanar subgraph G_2 such that $R(G_2) < R(G_1)$, etc. There is thus obtained a descending, necessarily finite chain $G, G_1,\ldots,G_k$ of nonplanar graphs such that G_{h-1} is a subgraph of G_h $(G_0 = G)$ and that $R(G_{h-1}) < R(G_h)$. The graph G_k will have the property that if H is any subgraph of G_k and $R(H) < R(G_k)$ then H is planar. It is therefore sufficient to assume that already.

(4.7) The nonplanar graph G is such that if H is any subgraph of G such that $R(H) < R(G)$ then H is planar.

According to Lemma (4.6) G contains two loops λ, λ' with exactly one common closed arc $\bar{\mu}$. Hence $\lambda + \lambda' - \mu$ (μ is the open arc) is a loop C of G. But if $H = G - \mu$ then $R(H) < R(G)$. Hence H is planar and the Jordan curve C is a subgraph of H. Assume already that H coincides with its representation in a plane Π. In Figure 41, C is the circle. Now no arc ν in $\Pi - H$ joins a to b of Figure 41. For if such a ν existed $H + \nu$ would be a representation of G in Π, and so G would be planar.

Let Ω be the finite region of Π (interior of C).

We face now two distinct possibilities according to the presence or absence in the graph H of an arc joining the two points a, b in Ω.

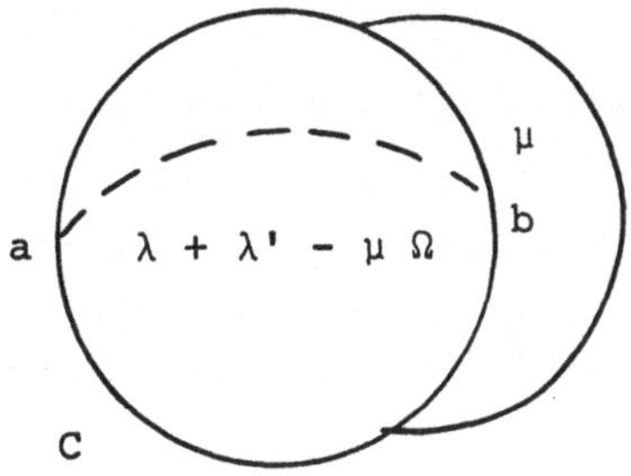

Figure 41.

These two situations must be examined separately.

I. <u>The points</u> a,b <u>are not joined by any arc of the graph</u> H <u>in</u> Ω. Since no arc may join a to b in Π - H, H must contain the only disposition corresponding to a nonplanar graph I itself, or G contains a subgraph such as I.

II. <u>The points</u> a,b <u>are joined by an arc of</u> H <u>in the region</u> Ω. Since no arc may join a to b in Π - H the only possibility is the one corresponding to II. This completes the proof of the theorem.

5. Reciprocal Networks

The general concept of reciprocal networks is this: two networks N,N^* are reciprocal if it is possible to turn a current and voltage distribution in N into a voltage and current distribution in N^* and conversely. It is evident that the graphs G and G^* of the two networks must be related in some special way. A more precise definition of reciprocity is that the following conditions must be fulfilled:

(a) There is a one-one correspondence between their branches. The branches may then be so numbered that this correspondence is actually $b_h \longleftrightarrow b_h^*$.

(b) Let i_h, v_h and i_h^*, v_h^* be current and voltage distributions in N and N^*. Then v_h is a distribution and i_h is a v_h^*, and conversely from N^* to N.

Let n_h and n_j^* denote the nodes of G and G^*. Kirchoff's current law for N must become his second law for N^*. Hence there must be a set of loops $\lambda_1^*, \ldots, \lambda_{\alpha_0}^*$ in G^* in one-one correspondence with the nodes $n_1, n_2, \ldots, n_{\alpha_0}$ of G_1 and likewise a set of loops $\lambda_1, \ldots, \lambda_{\alpha_0^*}$ in G in one-one correspondence with the nodes $n_1^*, \ldots, n_{\alpha_0^*}^*$ of G^*. This suggests at once the passage to certain complexes.

Let then K and K^* be the complexes obtained from G by means of cells e_h bounded by the λ_h, and similarly for K^* and the λ_j^*. This strongly suggests a reversal of the incidence relations between K and K^*. That is if $\delta b_h = n_k - n_j$ then in K^* the branch b_h^* is to be adjacent to the cells e_k^*, e_j^* that is to exactly two cells. Likewise if b_h^* is on the boundary of e_k^* then n_k must be an endpoint of b_h. Furthermore if $\delta e_k^* = \lambda_k^*$ then n_k must possess exactly one umbrella. In other words, except for connectedness K and K^* have all the surface properties (A,B of Chapter VIII, Section 1). We impose therefore the condition <u>that K and K^* are a complex defining a surface S and the dual of that complex</u>.

6. Duality of Electrical Networks

Let N and N_1 be two electrical networks and G and G_1 their graphs. As network-graphs one may assume at all events that both are connected.

Let the usual designation n_k, b_j prevail for G and denote by

n_{1k}, b_{1j} those appropriate for G_1. One would like to bring the branches b_j, b_{1j} into a one-one relation such that the current distributions in N become voltage distributions in N_1 and conversely. That is one would like to be able to interchange cycles and coboundaries in the two graphs G, G_1.

Let the branches be so numbered that their one-one correspondence is expressed by $b_j \longleftrightarrow b_{1j}$. The node conditions for a cycle at n_k become in G_1 loop conditions for a loop λ_{1k} of G_1 and there is a one-one correspondence $n_k \longleftrightarrow \lambda_{1k}$. Similarly there is a select collection of loops λ_j of G with a one-one correspondence $n_{1k} \longleftrightarrow \lambda_j$. This suggests immediately covering λ_{1k} and λ_j with cells e_{1k} and e_j, yielding as in Chapter VII, two complexes K_1 and K with $G = K - \Sigma\, e_j$, $G_1 = K_1 - \Sigma\, e_{1k}$.

(6.1) <u>Theorem</u>. <u>The complexes</u> K, K_1 <u>define two dual orientable surfaces</u>.

Since b_j has two endpoints n_h, n_k the branch b_{1j} belongs to exactly two loops $\lambda_{1h}, \lambda_{1k}$ and hence b_j is adjacent to exactly two cells e_{1h}, e_{1k} of K_1. Thus the surface property A of Chapter VII, Section 1 holds for K_1, and similarly also for K. From the fact that δe_{1k} consists of just one loop we also conclude that n_k possesses the umbrella property B and similarly for K. Hence, except for connectedness of K , K and K_1 are surfaces. However K is connected. Hence any two nodes of K may be joined by an arc of G. Hence any pair of cells of K_1 may be joined by a geometric chain in K_1: K_1 is connected. Thus K and K_1 are surfaces. The nature of the correspondence between their elements shows that they are duals.

Let now the branches and cells be oriented in K and let them be oriented in K_1 so that if η is the incidence matrix $b_h \to n_k$

in K then η' is the incidence matrix $e_{1h} \to b_{1k}$ in K_1. In any row of η two elements are $+1,-1$ and the rest zero. Hence this holds also in any column of η'. Therefore

$$\delta \, \Sigma \, e_{1h} = 0.$$

$$\gamma = \Sigma \, e_{1h}$$

is a 2-cycle of K_1. Hence K_1 is orientable. From

$$R_2(K) = R_2(K_1) = 1$$

follows that K is likewise orientable. This completes the proof of the theorem.

The problem of <u>network duality</u> is now easily solved. In fact:

(6.2) <u>Theorem</u>. <u>N.a.s.c. in order that the two networks</u> N <u>and</u> N_1 <u>be dual is that their graphs</u> G, G_1 <u>be spherical (= planar)</u>. <u>If say the complex</u> K <u>of</u> G <u>is taken as the sphere then</u> G <u>and</u> G_1 <u>are dual in the sphere</u>.

For duality between the two networks implies that the cycles of G represent the voltage of N_1. Moreover, "coboundary" now means "boundary in the sphere". Hence the cycles of K all bound in K and so $R_1(K) = 0$, that is (Section 1) K is a sphere. Then also the cycles of K_1 bound in K_1 and hence they represent the voltages of N. Since "cycle" and "current distribution" are interchangeable terms the theorem is proved.

PART II

THE PICARD-LEFSCHETZ THEORY AND FEYNMAN INTEGRALS

PREFACE

I learned very recently of an important mathematical connection between old work of Picard and myself and the theory of Feynman integrals. A central question in particle physics, mathematicians seem to know very little about it. I believe that some elementary mathematical analysis together with my early topological work, might serve to attract the attention of some younger physicists and mathematicians to this general problem. The following short monograph has resulted from this conviction.

As my "topology" belongs to the first third of this Century it has to be the basis of my work.

The monograph consists of the following main parts: I. Algebraic and topological résumé, with almost no proofs. II and III. The Picard-Lefschetz theory and extensions. IV. Feynman integrals.

The profuse mathematical contributions in the physical literature are based almost entirely upon very up to date differential topology. Since the Feynman theory comprises only algebraic structures, differential topology seemed really excessive. As I am an old hand at algebraic geometry and topology, both together seemed to be quite sufficient for early starters. I hope that the following pages go some way toward proving these assertions.

I have derived much profit from reading Poincaré's paper [8] on residues, and its enormous extension by Leray [4]. Much profit was derived from reading Hwa-Teplitz [1] and Pham [5]. It is also with pleasure that I recall that Professors Regge and Wightman listened with patience to long discourses of mine, and that both of them warmly encouraged me to pursue my present task.

July 1971 Solomon Lefschetz

INTRODUCTION

Let the equation

$$F(x,y,z) = 0$$

represent a complex irreducible algebraic surface. It is assumed that the surface has no other singularities than a general projection from a complex projective 4-space. Moreover, although the representation is cartesian, the surface is really assumed to be in a complex projective 3-space with infinity taken care of by a projection of type

$$x = \frac{1}{x'},\ y = \frac{y'}{x'},\ z = \frac{z'}{x'} .$$

However, from the point of view of the problem of interest, this is not too important.

Let $\{H_z\}$ denote the sections of the surface by the pencil of planes $z = \text{const}$. Let the plane $z = 0$ be tangent to F at the origin and this so that H_0 has the origin as mere double point with distinct tangents. Consider now an abelian integral

$$J(z) = \int_\gamma R(x,y,z)dx$$

taken along a path γ in H_z. Suppose that while the integral is well behaved away from $z = 0$ the path γ tends to pass through the origin as $z \to 0$. As part of his fundamental study of simple and double integrals of rational functions taken on F, Picard determined exactly the behavior of the integral $J(z)$ near the origin by studying the variation of γ as z turned around $z = 0$. Some fifteen years later I gave an exact topological determination for all possible paths. I also dealt with relatively simple extensions. All told this constitutes the theorem of Picard-Lefschetz.

Coming now from a different order but related idea, some two decades ago Feynman in the course of investigations on particle physics introduced multiple integrals in whose description I shall utilize a general notation repeatedly occurring throughout this paper. Let $(\alpha_1,\ldots,\alpha_p)$ be a finite collection of symbols whose range is known from the context. We shall use the quasi-vector designation $\underline{\alpha}$ to describe the collection.

Set then $\underline{x} = \{x_1,\ldots,x_n\}$, $\underline{y} = \{y_1,\ldots,y_n\}$ where the x_h are real or complex coordinates and the y_k likewise real or complex parameters. Let $Q_h(x,y)$, $0 < h \leq s$ denote real quadratic polynomials and let $d\underline{x} = dx_1 \cdots dx_n$ be skew symmetric products. The Feynman problem consists in the study of the analytical character as function of $\underline{y}$, of

$$J(\underline{y}) = \int_\Gamma \frac{d\underline{x}}{\prod_h Q_h(\underline{x},\underline{y})}$$

where Γ is the whole admissible part of the $\underline{x}$ space. Admissible since Γ may not cross the sets $Q_h = 0$. One must therefore deform Γ around the singular loci in the complex extension of the space $\underline{x}$ and describe the consequences, The analogy with the Picard problem is clear.

In essence the problem reduces to the topological study of the various possibilities of by-passing the different types of singularities that may well arise. That is, one must study the possible topologies around each singularity. Since the simplest singularity is an isolated point much space will first be devoted to this type and various extensions will be dealt with later.

My purpose in the following pages is to present an introductory treatment suggested by the general problem which combines the Picard and Feynman problems.

The major part of the treatment must consist in the manner of bypassing all singularities, infinity included. This is a largely topological question. The topological foundations have been completely established in my book Topology [2]. I will present à résumé of the required material.

Some standard mathematical symbols. We admit acquaintance with symbols $\cup$, $\cap$, $A-B$, $\supset$, $\subset$, $\in$. Less well known are $\{x|\cdots\}$ = all elements with property $\cdots$. Thus $\{x_h | 1 \leq h \leq n\}$ = the collection $(x_1, x_2, \ldots, x_n)$.

A matrix with terms a_{jk} (j for rows, k for columns) is written $[a_{jk}]$.

P stands for Euclidean product.

K generally denotes a finite simplicial complex and σ are its simplexes.

M^n represents a real compact orientable n-manifold.

V^d stands for "complex algebraic d-dimensional variety".

$\mathcal{P}^n$ represents a complex projective n-space. When referred to complex projective coordinates $x_o, x_1, \ldots, x_n$ then $\mathcal{P}^n_h$ represents the 2n cell $x_h \neq 0$. In particular $\mathcal{P}^n_o$ has the projective coordinates $1, x_1, \ldots, x_n$. We then (carelessly) consider the x_h as the complex coordinates of $\mathcal{P}^n_o$.

Some useful designations related to analytic functions of several complex variables. Let $x_1, x_2, \ldots, x_n$ be cartesian coordinates for a complex n-space. Let 0 denote the origin. We are particularly concerned with functions $F(\underline{x})$ holomorphic at the origin. Let $F(0) = \alpha$. We refer to F as:

a nonunit when $\alpha = 0$,

a unit when $\alpha \neq 0$.

If F is a unit so is F^{-1} (but with a possibly smaller

convergence region). $E(\underline{x})$ is the general designation for a unit.

Recall these basic theorems of Weierstrass.

<u>Preparation theorem</u>. <u>Suppose that</u> $F(x_1,0,\ldots,0) = x_1^p E(x_1)$. <u>Then</u>

$$E(\underline{x})F(\underline{x}) = x_1^p + a_1(x_2,\ldots,x_n)x_1^{p-1} + \cdots + a_p(x_2,\ldots,x_n)$$

<u>where the</u> a_h <u>are nonunits</u>.

The polynomial in x_1 is said to be special. The integer p, the least degree of a term in x_1 alone, is the degree of F in x_1.

A nonunit $F(\underline{x})$ is <u>irreducible</u> when one may not write $F \equiv GH$ where G and H are both nonunits.

<u>Factorization theorem</u>. <u>Any nonunit may be decomposed into nonunit factors</u>

$$F \equiv F_1^{n_1} F_2^{n_2} \ldots F_\gamma^{n_\gamma}$$

<u>where the</u> F_h <u>are irreducible and unique (except for their order) to within unit factors</u>.

<u>Topological concepts</u>. All the spaces considered will be topologically identical with Euclidean subsets. The standard concepts related to such subsets are assumed familiar to the reader. We emphasize particularly the fundamental notion of compactness and recall that:

> an n-cell E^n is the topological image of an open spherical region of Euclidean n-space;
>
> a topological (n-1)-sphere, $n > 1$, is the topological image of the sphere of Euclidean n-space.

Recall also these designations: if a,b, $a < b$, are real numbers then (a,b) denotes the interval (one-cell) $a < x < b$; $[a,b]$

denotes the segment $a \leq x \leq b$; $[a,b)$ or $(a,b]$ denote the sets $a \leq x < b$ or $a < x \leq b$.

If V denotes a real variety a simplicial covering complex — if it has one — will be denoted by V_*.

CHAPTER I

TOPOLOGICAL AND ALGEBRAIC CONSIDERATIONS

1. Complex Analytic and Projective Spaces

A complex analytic 2n-manifold M^{2n} is a connected space defined by the following properties:

(1.1) *It has a finite covering by 2n-cells* U,V,...,W *each parametrized by* n *complex coordinates respectively* $u_h, v_j, \ldots, w_k$ *with metrics defined by*

$$d_u^2 = \sum u_h \bar{u}_h, \ldots .$$

(1.2) *At all points of overlap say of* U,V, *the* v_k *are holomorphic in the* u_h.

If A is a fixed point of U V and Q a variable point of U V which → A then (1.2) implies that as one of the distances $d_u(A,Q)$, $d_v(A,Q) \to 0$ so does the other. That is the various distances are mutually coherent. Hence the set of all spherical regions on every cell U,..., is a suitable open set base for an open set topology of our space and is the one adopted henceforth.

(1.3) *Orientation*. Let $u_h = u_h' + iu_h''$. Orient U by naming the real coordinates in the order $u_1', u_1'', u_2', \ldots, u_n''$. By standard methods one may show that this provides an orientation for M^{2n}. It is thus shown to be an orientable manifold.

2. Application to Complex Projective n-space $\mathcal{P}^n$

It is defined by the following properties:

I. To each point of $\mathcal{P}^n$ there corresponds an ordered collection of $n + 1$ complex numbers $x_o, x_1, \ldots, x_n$ not all zero.

The collections $\{x_j\}$ and $\{kx_j\}$, $k \neq 0$, and only these define the same point.

The x are the projective coordinates of their point.

II. A complex transformation of coordinates

$$x'_i = \sum a_{ij}x_j, \quad |a_{ij}| \neq 0, \tag{2.1}$$

merely sets up a new correspondence of $\mathscr{P}^n \leftrightarrow \{x_j\}$ but does not change the space $\mathscr{P}^n$.

(2.2) Theorem. One may assign to $\mathscr{P}^n$ an open set topology under which it is a complex analytic M^{2n}. Moreover under this topology the transformation (2.1) is topological.

On $\mathscr{P}^n_h(x_h \neq 0)$ one may take $x_h = 1$, that is coordinates $X_1, X_2, \ldots, X_{h-1}, 1, X_{h+1}, \ldots, X_n$. Upon assigning to $\mathscr{P}^n_h$ a metric defined by

$$d_h^2 = \sum X_j \bar{X}_j \quad (j \neq h)$$

it becomes an analytic 2n-cell parametrized by the X_h. Thus $\mathscr{P}^n$ has a finite covering by analytic 2n-cells. All that is needed, therefore is to show that wherever cells overlap their parameters are mutually holomorphic. This only needs to be done for any two of the pairs of cells say $\mathscr{P}^n_o$ and $\mathscr{P}^n_1$. Let their coordinates be X_n for $\mathscr{P}^n_o$ and Y_h for $\mathscr{P}^n_1$. Since projective coordinates are proportional at a point Q of $\Omega = \mathscr{P}^n_o \cap \mathscr{P}^n_1$ we have

$$\frac{1}{Y_o} = \frac{X_1}{1} = \frac{X_h}{Y_h}, \quad h \geq 2.$$

Hence

$$Y_o = \frac{1}{X_1}, \quad Y_h = \frac{X_h}{X_1}, \quad h \geq 2.$$

Since on Ω: $X_o \neq 0$, $X_1 \neq 0$, the Y_h are everywhere holomorphic in the X_h. The converse is obvious. As we have seen this suffices to define an adequate open set topology for $\mathcal{P}^n$.

Passing now to the topological nature of (2.1) if $\mathcal{P}_h'^n$ is the set $x_h' \neq 0$ it is sufficient to show concordance of the topologies of $\mathcal{P}_h^n$ and $\mathcal{P}_h'^n$ whenever they overlap. This may be safely left to the reader.

(2.3) Remark. It is easily shown that the adopted $\mathcal{P}^n$ topology induces on a subspace $x_h = 0$ the analogous $\mathcal{P}^{n-1}$ topology.

To prove that $\mathcal{P}^n$ is a complex analytic M^{2n} it suffices to show that it has these two properties:

The space $\mathcal{P}^n$ is connected. The $\mathcal{P}_h^n$ are cells hence connected. Since they have the common point $(1,1,\ldots,1)$ $\mathcal{P}^n$ is also connected.

(2.4) The space $\mathcal{P}^n$ is compact. This follows from the fact that $\mathcal{P}^n$ may be covered by a finite simplicial complex.

For the proof see [10], p. 133.

3. Algebraic Varieties

Hereafter, all polynomials or forms (homogeneous polynomials) are assumed with complex coefficients.

Let $\mathcal{P}^n$ and $\underline{x} = \{x_o, x_1, \ldots, x_n\}$ be a complex projective n-space and its coordinates. Let $F(\underline{x})$ be a form. The set of points of $\mathcal{P}^n$ defined by the relation

$$F(\underline{x}) = 0 \tag{3.1}$$

is an algebraic hypersurface of $\mathcal{P}^n$. An algebraic variety V is merely the intersection of a finite set of hypersurfaces, that is the set of points satisfying a finite system

$$F_h(\underline{x}) = 0, \quad 1 \leq h \leq s. \tag{3.2}$$

The variety V is <u>irreducible</u> whenever if G,H are two forms and GH = 0 at all points of V then one of the factors say G = 0 at all points of V. For a hypersurface irreducibility in this sense is the same as in the algebraic sense.

Let V be irreducible and agree to set $G \equiv 0$ whenever it vanishes at all points of V. Denote such forms by G^*. Define a <u>rational function</u> R as the quotient $\frac{H}{G}$ of two forms of equal degree. Define now two rational functions

$$R = \frac{H}{G}, \quad R_1 = \frac{H_1}{G_1}$$

where G and G_1 are not a G^*, as <u>identical</u> whenever

$$R - R_1 = \frac{K^*}{GG_1} .$$

Let R^* denote a class of rational functions thus identified. The collection $\{R^*\}$, with elements combined like "natural" numbers constitutes a <u>field</u>: the <u>function field</u> C(V) of the variety V. The maximum number d of algebraically independent elements of C(V) is the complex <u>dimension</u> of V. One refers to V as a <u>d-variety</u> and often denotes it by V^d (2d = the classical Menger-Urysohn dimension of V^d).

(3.3) <u>Algebraic varieties are compact</u>. For they are closed in a $\mathcal{P}^n$ which is compact.

(3.4) <u>Generic points and varieties (van der Waerden)</u>. Let V^d be an irreducible variety defined say by (3.2). A point $\underline{\xi}$ of V^d is <u>generic</u> for V^d whenever it satisfies no other algebraic relations than (3.2) or their consequences.

The term "generic" is also used in a (seemingly) wider sense described in the following example. Let $f(\underline{a},x_o,x_1,x_2) = 0$ be the equation of a family of conics depending upon a set of six homogeneous coefficients $\underline{a} = \{a_k | 0 \leq k \leq 5\}$ and satisfying an irreducible system $\underline{\phi}(\underline{a}) = 0$. Thus every solution $\underline{\alpha}$ of this system defines a unique conic $\gamma_{\underline{\alpha}}$ of the family. We call $\gamma_{\underline{\alpha}}$ a generic curve of the family $\{\gamma\}$ whenever $\underline{\alpha}$ is a generic point of $\phi(\underline{a}) = 0$. The extension to irreducible families of varieties is clear.

(3.5) Algebraic varieties in cartesian coordinates. Very frequently, especially in analytic considerations we shall be mainly interested in the part of varieties situated say in $\mathscr{P}^n_s$, that is in the cell $x_s \neq 0$. We may then set $x_s = 1$, and utilize cartesian coordinates $\{X_h | 1 \leq h \leq n\}$. One may then consider the variety as defined by a hypersurface

$$F(\underline{X}) = 0$$

where F is now a complex polynomial or more generally by the intersection of a finite set of hypersurfaces:

$$F_h(\underline{X}) = 0, \quad 1 \leq h \leq s.$$

The previous definitions of irreducibility, etc., apply in toto save that rational functions are now merely quotients of any polynomials (denominators still $\neq 0$ on V^d).

(3.6) Singularities of varieties. Let V^d be the same irreducible complex variety as before defined by

$$F_k(x_o,\ldots,x_n) = 0.$$

A point $\underline{\xi}$ of V^d is nonsingular whenever the system in the dx_h at $\underline{\xi}$

$$\frac{\partial F_k(\underline{\xi})}{\partial \xi_h} \cdot dx_h = 0 \tag{3.7}$$

has a Jacobian matrix

$$\left[\frac{\partial F_k(\underline{\xi})}{\partial \xi_h}\right]$$

of rank $n - d$ at $\underline{\xi}$.

Let $\underline{M}(x) = \{M_s(x)\}$ be the set of minors of order $n - d$ of the matrix $[\frac{\partial F}{\partial x}]$. The system $\underline{F}(\underline{x}) = 0$, $\underline{M}(\underline{x}) = 0$ is of rank $n - d$, while if it is of lower rank the point $\underline{\xi}$ is <u>singular</u>. The system (3.7) defines the singular variety W of V^d. It is covered by a complex W_* of the one covering V^d. Since $\dim W \leq d - 1$, $V^d - W$ is connected. Thus $V^d - W$ is covered by a locally finite collection of analytic 2d-cells $[U_\alpha]$. We call it an open analytic 2d-manifold M^{2d}.

(3.8) <u>General remark</u>. We shall repeatedly shift to the real domain, but only in arguments applicable to both real and complex domains.

4. A Résumé of Standard Notions of Algebraic Topology

More detailed information may be found in my three books (two in Colloquium Lectures, the earliest also reprinted by Chelsea).

(4.1) A p-<u>simplex</u> σ^p (dimension always upper index) is a collection of Euclidean points or vectors given by

$$A = \sum_{h=0}^{p} x_h A_h, \quad 0 < x_h < 1, \ \sum x_h = 1.$$

Replacing $p - q$ of the x_h by zero yields a q-face σ^q of σ^p; the union σ^q, $(q < p)$ is the boundary $\partial\sigma^p$ of σ^p; $\sigma^p \cup \partial\sigma^p = \text{Cl}\ \sigma^p$: <u>closure</u> of σ^p.

A p-<u>complex</u> K^p is a collection $\{\sigma^q_j \mid 0 \leq q \leq p;\ 1 \leq j \leq \alpha_q\}$ such that $\sigma \in K \Rightarrow \partial\sigma \in K$; any two σ's are disjoint.

One orients σ^q by naming its vertices A_j in a definite order modulo an even permutation. An odd permutation replaces σ^q by $-\sigma^q$. If $q = 0$: A_j one merely writes $+A_j$ or $-A_j$; hereafter all simplexes are oriented.

A q-chain c^q of K over $\mathcal{G}$ = a field or the set of integers (or quaternions in IV) is a formal expression

$$c^q = \sum x_j \sigma_j^q, \quad x_j \in \mathcal{G}.$$

If, say

$$\sigma^q = (A_o, \ldots, A_q) \in K$$

its chain boundary or merely boundary $\partial\sigma^q$ is

$$\partial\sigma^q = \sum_r (-1)^r A_o \ldots A_{r-1} A_{r+1} \ldots A_q$$

and also if c^q is as above

$$\partial c^q = \sum x_j \partial\sigma_j^q .$$

If $\partial c^q = 0$ one calls c^q = q-cycle of K over $\mathcal{G}$ (of K omitted when clear from contex).

One verifies that $\partial\partial\sigma^p = 0$, hence also $\partial\partial c^p = 0$: ∂c^p is a cycle. The collection $\{c^p \mid c^p = \partial c^{p+1}\}$ is denoted by F^p. We also write $c^p \sim 0$: read c^p is a bounding cycle.

If Z^p is the collection of all p-cycles over $\mathcal{G}$ then $F^p \subset Z^p$. Hence

$$H^p = Z^p/F^p$$

(factor group) is an additive group: the pth homology group over ($\sim$ is the symbol of "homology"). The number $R^p = \dim H^p$ (over integers or a field) is the pth Betti number of K over $\mathcal{G}$.

Let α_q be the number of σ^q of K. Then the characteristic

$$\chi(K) = \sum(-1)^q R^q = \sum(-1)^q \alpha_q.$$

(4.2) Relative notions. A subcomplex L of K is a subset $\{\sigma^q\}$ of its simplexes which (a) is a complex; (b) if $\sigma \in K$ has all vertices in L then $\sigma \in L$. One calls L a closed subcomplex of K. Then naturally, $K - L = L_1$ is called an open subcomplex of K.

Interesting exercise: Define "open" and "closed" complexes without reference to K.

Actually only the notions of closed and open subcomplexes will occur in the sequel.

Returning then to L = closed subcomplex of K, we define a chain c^q of K with $\partial c^q \subset L$ as a relative q-cycles of K or cycle mod L of K. We call $c^q \sim 0 \bmod L$ (= bounding mod L) whenever

$$c^q = \partial c^{q+1} + d^q, \quad d^q \subset L.$$

To the cycles mod L one may extend all the properties of ordinary cycles.

The star $\operatorname{St}\sigma$ of $\sigma \in K$ is the set of all $\sigma' \in K$ with $\partial\sigma$ (including σ) as a face.

(4.3) Still assume the simplexes of K rectilinear in some Euclidean n-space $\mathscr{E}^n$. The metric of the latter is extended to K. In particular

$$\text{Mesh } K = \sup \operatorname{diam} \sigma \,|\, \sigma \in K.$$

If $\Sigma = \{\sigma'\}$ is any collection of simplexes of K, denote by $|\Sigma|$ the set of all points of the σ'. In particular $|K|$ is called a polyhedron.

(4.4) Manifolds. The connected complex K is an absolute n-manifold [n-manifold mod L] whenever $|\operatorname{St}\sigma|$, $\sigma \in K$, $[\sigma \in K - L]$ is an n-cell.

Let M^n be an <u>absolute</u> n-manifold. It has then an n-cycle — the union of its σ^n — which is an absolute n-cycle γ^n, or merely one mod 2. In the first case M^n is <u>orientable</u>, in the second <u>non-orientable</u>. If orientable, M^n may be oriented by γ^n or $-\gamma^n$ (two opposite orientations).

Same thing for the relative M^n.

(4.4a) <u>Duality</u>. According to Poincaré for an absolute M^n:

$$R^p = R^{n-p}.$$

I have also shown that for an M^n mod L

$$R^p(M^n-L) = R^{n-p}(M^n,L). \tag{4.4b}$$

Moreover, if L_1,L_2 are two closed subcomplexes of an absolute M^n then

$$R^p(M^n-L_1,L_2) = R^{n-p}(M^n-L_2,L_1). \tag{4.4c}$$

(4.5) <u>Subdivision</u>. Denote by $\hat{\sigma}$ the barycenter of σ. Define the subdivision operation D recursively as follows: $D\sigma^o = \hat{\sigma}^o = \sigma^o$. If σ_i^{p-1} are the faces of σ^p define

$$D\sigma^p = \bigcup(\hat{\sigma}^p,D\sigma_i^{p-1}) \cup D\sigma_i^{p-1} \cup \hat{\sigma}^p.$$

In an obvious sense this expression may be formulated as

$$D\sigma^p = D\partial\sigma^p \cup (\hat{\sigma}^p,D\partial\sigma^p).$$

Thus D is defined for all K. It results in a new complex $K' = DK$: the <u>derived</u> of K. The successive derived are then

$$K^{(n)} = DK^{(n-1)}.$$

Any derived of K is called <u>subdivision</u> of K.

One proves readily that

$$\partial c^{q+1} = c^q \Longrightarrow \partial Dc^{q+1} = Dc^q, \quad \text{all } q.$$

Hence:

(4.6) Subdivision does not alter (a) the homology and Betti numbers absolute or relative; (b) the manifold properties.

5. Homotopy. Simplicial Mappings

(5.1) As applied to Euclidean spaces homotopy means this: Given two Euclidean sets X, Y and two mappings $\phi, \psi\colon X \to Y$, they are homotopic whenever there is a third mapping ω of $X \times \ell$ $(\ell = [0,1])$ into Y such that $\phi = \omega|(X \times 0)$ and $\psi = \omega|(X \times 1)$.

Necessary conditions for this homotopy: if $x \in X$ and $y' = \phi(x)$, $y'' = \psi(x)$ then a closed segment joins y' and y'' in Y.

(5.2) Simplicial mappings. Let $K = \{\sigma\}$, $L = \{\zeta\}$ be two simplicial complexes (the ζ's are the simplexes of L). A mapping $\mu\colon L \to K$ is simplicial whenever every $\mu\zeta$ is a simplex of K.

(5.3) Theorem. Every mapping $\mu\colon L \to K$ may be ε-approximated by a simplicial mapping of some subdivision of L into one of K (Alexander).

(5.4) A fundamental generalization. Let K be simplicial with cells e_h^p. Let μ be a topological mapping of $|K|$ into some space X. The formal interrelations of the image cells $\mu e_h^p = g_h^p$ are identical with those between the e_h^p. We consider $\{g_h^p\}$ as a complex $\mathscr{K}$ and shall extend to it automatically all the terminology utilized for K.

6. Singular Theory

Let σ^p be a p-simplex, X a compact metric space, f a mapping $\bar{\sigma}^p \to X$. The pair (σ^p, f) is a singular p-simplex in X. If σ'^p is another p-simplex and ψ is a linear homeomorphism $\bar{\sigma}'^p \to \bar{\sigma}^p$ we agree that the two singular p-simplexes (σ^p, f) and $(\sigma'^p, f\psi)$ are identical. If σ^q is an oriented q-face of σ^p then (σ^q, f) is a singular q-simplex called a q-face of (σ^p, f).

Singular chains are defined in the obvious way as finite sums

$$c^p = \sum m_h(\sigma_h^p, f_h). \tag{6.1}$$

If

$$\partial\sigma^p = \sum r_j \sigma_j^{p-1}$$

then by definition

$$\partial(\sigma^p, f) = \sum r_j(\sigma_j^{p-1}, f).$$

Hence, for the singular chain (6.1)

$$\partial c^p = \sum m_h \partial(\sigma_h^p, f_h) \sim 0, \qquad \partial\partial = 0,$$

and c^p is a singular p-cycle if $\partial c^p = 0$. Also ∂c^p is defined as a singular bounding (p-1)-cycle.

Let $|\sigma^p, f| = f\bar{\sigma}^p$. This is a compact set, the carrier of (σ^p, f). For the chain (6.1) define its carrier $|c^p|$ as the union, which is compact, of the carriers $|\sigma_h^p, f_h|$, for all $m_h \neq 0$. Clearly $|\partial c^p| \subset |c^p|$.

Let A be a compact subset of X. If the singular chain c^p has its boundary $\partial c^p \subset A$ define it as a singular p-cycle mod A. We also call c^p "singular bounding" and write $c^p \sim 0 \bmod A$ whenever there is a singular chain c^{p+1} such that $\partial c^{p+1} = c^p + d^p$, $|d^p| \subset A$.

A singular c^p is defined as in $X - A$ if its carrier

$|c^p| \subset X - A$. This gives rise automatically to the boundary and related relations for $X - A$.

(6.2) Theorem. Let $K = \{\sigma\}$. If we consider the $\{(\sigma,$ identity$)\}$ as singular simplexes then the homology groups of the singular simplexes of $|K|$ are isomorphic with those obtained from the singular cells (σ, identity) that is from those of K itself. Hence the homology groups of K are topological invariants. Similarly if L is a closed subcomplex of K and the homology groups of K mod L and of $K - L$.

(6.3) Applications. The Betti numbers of an n-sphere $n > 0$, are $R^o = R^n = 1$, $R^k = 0$, $k \neq 0, n$. The Betti numbers of a p-cell, $p \neq 0$, are $R^o = 1$, $R^q = 0$ for $q > 0$.

(6.4) Subdivision of a singular cell (σ, f) is defined by reference to the antecedent σ.

7. The Poincaré Group of Paths

Let X be an arcwise connected space. Let A be a point of X and consider the image γ in X of an oriented segment ℓ whose two ends are imaged into A. Let γ' be a similar image of an oriented segment ℓ'. Then γ followed by γ' is again an image γ'' of a similar segment ℓ'' with both end points imaged into A, and we define $\gamma'' = \gamma'\gamma$. If γ is homotopic to A in X we define $\gamma = 1$. Finally, the image of ℓ inverted defines γ^{-1}. Under these definitions there is obtained a group $G(A)$. If A is replaced by another point B one shows readily that $G(B) \cong G(A)$, under the operation $\gamma \to \lambda^{-1}\gamma\lambda$ where λ is a directed arc BA. The abstract group thus defined, manifestly a topological invariant of X, is known as group of the paths, or Poincaré group of X, and designated by $\pi(X)$.

(7.1) The group $\pi(K)$ of a simplicial complex K is a factor group of a free group on a finite number of generators.

We may evidently suppose that K is connected. Let K_2 be the subcomplex of K consisting of all closed 2-simplexes (triangles) of K. Let also Γ be the graph consisting of the closed arcs of K. Evidently, both K_2 and Γ are connected. And clearly $\pi(K) = \pi(K_2)$. We may, therefore, assume that $\dim K = 2$, that is that $K = K_2$.

Let now $\{\lambda_1, \ldots, \lambda_R\}$ be a maximal set of independent loops of Γ as one-cycles of Γ. Thus $R = R_1(\Gamma)$. Take a fixed vertex A of Γ and join it by an arc $\mu_k = AA_k$ to a vertex A_k of λ_k. Thus $\mu_k^{-1}\lambda_k\mu_k = \lambda_k^*$ is an operation of $\pi(\Gamma)$ based on A and $\{\lambda_k^*\}$ is manifestly a set of independent generators for $\pi(\Gamma)$. Now in K the λ_k^* may be subject to a certain number of relations, resulting from the possible ∂c_2 among the chains $\Sigma\, m_k\lambda_k^*$. These relations generate a subgroup G of $\pi(\Gamma)$ and $\pi(K) = \pi(\Gamma)/G$ (factor group).

8. Intersection Properties for Orientable M^{2n} Complex

(8.1) Fundamental theorem. Let $c^p, c^q, p + q \geq 2n$, be chains of $K - L$ (orientable M^{2n}) whose boundaries $\partial c^p, \partial c^q$ are disjoint (their carriers are disjoint). Then one may define their intersection as a $(p+q-2n)$-chain of $K' - L'$, written $c^p \cdot c^q$ and it satisfies the relation

$$\partial(c^p \cdot c^q) = c^p \cdot \partial c^q + (-1)^{2n-q}\partial c^p \cdot c^q.$$

If c^p and c^q are disjoint $c^p \cdot c^q = 0$. Hence if each chain is disjoint from the boundary of the other their intersection is a $(p+q-2n)$-cycle.

(8.2) Corollary. If say $c^p = \gamma^p$, a cycle, then

$$\partial(\gamma^p \cdot c^q) = \gamma^p \cdot \partial c^q. \tag{8.3}$$

Hence if γ^p and γ^q are cycles of $K - L$ their intersection $\gamma^p.\gamma^q$ is a $(p+q-2n)$-cycle of $K' - L'$.

Finally let c^p and c^q be singular chains of $K - L$: their boundary-carriers are disjoint from $\overline{St\ L}$. Then by arbitrary approximations by chains of $K^{(m)} - L^{(m)}$ one may show that the various singular chains considered may be arbitrarily closely approximated by chains of some $K^{(m)} - L^{(m)}$, and hence that the various intersections considered have topological character. Explicitly:

(8.4) All the above properties continue to hold for singular chains and cycles of $K - L$.

(8.5) Special case $p + q = 2n$. Then $c^p.c^{2n-p}$ with non-intersecting boundaries is a cycle γ^o and so it has a Kronecker index also called intersection number which we denote more simply by (c^p,c^{2n-p}) and we have

$$(c^p,c^{2n-p}) = (-1)^p(c^{2n-p},c^p).$$

Finally from (8.2)

$$\partial(c^{p+1}.\gamma^{2n-p}) = (-1)^p(\partial c^{p+1}.\gamma^{2n-p}).$$

Hence if $\gamma^p = \partial c^{p+1}$, that is if $\gamma^p \sim 0$ then

$$(\gamma^p,\gamma^{2n-p}) = 0$$

whatever γ^{2n-p}. Similarly with p and $2n-p$ interchanged. This is also obvious from

$$(\gamma^p,\gamma^{2n-p}) = (-1)^p(\gamma^{2n-p},\gamma^p).$$

All this applies likewise to appropriate singular chains. We note in particular that if c^p and c^{2n-p} are singular with disjoint boundaries then the index (c^p, c^{2n-p}) is uniquely defined and the properties just proved hold for it.

(8.6) <u>Application</u>. Let L_1, L_2 be normal subcomplexes of K such that $K - (L_1 \cup L_2)$ is a 2n-manifold. It is easily seen that everything just said holds if γ^p is a cycle of $K - L_1 \bmod L_2$ and δ^{2n-p} (in place of γ^{2n-p}) is a cycle of $K - L_2 \bmod L_1$. Then one may define the index $(\gamma^p, \delta^{2n-p})$ whatever γ^p and δ^{2n-p} and it depends solely upon the appropriate homology classes: of γ^p as a cycle of $K - L_1 \bmod L_2$ and of δ^{2n-p} as a cycle of $K - L_2 \bmod L_1$. Moreover, this index has topological character.

Remembering now that say the cycles of $K - L_1 \bmod L_2$ depend upon a finite number of independent relations between a finite set of oriented simplexes, we may state:

(8.7) <u>The homology groups</u> $H^p(K-L_i, L_j)$, $i \neq j$, $i,j = 1,2$ <u>have the same structure as those of a finite complex. In particular, the associated Betti numbers are finite.</u>

We also note this important property:

(8.8) <u>N.a.s.c. in order that some multiple of</u> $\gamma^p \sim 0$ <u>in</u> $K - L_1 \bmod L_2$ <u>is that</u>

$$(\gamma^p, \delta^{2n-p}) = 0$$

<u>for every</u> δ^{2n-p}, <u>and conversely. Hence if</u> $\{\gamma_h^p\}$ <u>and</u> $\{\delta_j^{2n-p}\}$ <u>are maximal independent sets of their types the determinant</u>

$$|(\gamma_h^p, \delta_j^{2n-p})| \neq 0.$$

9. Real Manifolds

For later purposes we shall have to consider <u>real</u> analogues of the preceding situation. The only modifications required in the statements (a) to replace everywhere "complex" by "real", and (b) to suppose that whenever two neighborhoods U_α and U_β overlap the related Jacobians

$$\left| \frac{\partial u_{\alpha j}}{\partial u_{\beta k}} \right| > 0.$$

We note explicitly that the only intersection that we will require are of type (H,c), where H is any hyperplane of V^n. Let H^* denote the M^{2n-2} part of H. Then: If γ^p is a cycle of $H^* \cap M^{2n}$ and γ^{2n-p} a cycle of M^{2n} then:

(a) $(\gamma^{2n-p},\gamma^p)_{M^{2n}} = ((H,\gamma^{2n-p}),\gamma^p)_{H^*}$;

(b) the two homologies $\gamma^{2n-p} \sim 0$ in M^{2n} and $(H,\gamma^{2n-p}) \sim 0$ in H^* are equivalent.

For details on covering complexes of analytic manifolds the reader may consult the paper of Lefschetz-Whitehead, Trans. Am. Math. Soc. 35, 310-316, 1933.

CHAPTER II

THE PICARD-LEFSCHETZ THEORY

1. Genesis of the Problem

This problem arose in the study, first undertaken by Émile Picard in the period 1880-1905, of the extension to algebraic surfaces of the classical Riemann theory. The more remarkable part of this work is that Picard carried it out with the scantiest aid of algebraic topology, since indeed that chapter of mathematics was entirely in its infancy.

We are particularly interested here in a special phase of Picard's program. Let

$$F(x,y,z) = 0 \tag{1.1}$$

be a complex irreducible algebraic surface whose sections $z = \text{const.}$ are denoted by H_z. Suppose that A is an ordinary point of F with tangent plane $z = z_o$ at the point and such that H_{z_o} has A as double point with distinct tangents. Then a suitable linear transformation

$$z \to z + z_o,\ x \to \alpha x + \beta y,\ y \to \gamma x + \delta y,\ \alpha\delta - \beta\gamma \neq 0$$

will reduce A to the origin 0 and the equation of the surface F about 0 to the form

$$F \equiv z + x^2 + y^2 + 2z(ax+by) + \cdots = 0,$$

the dots indicating here and later neglected terms of higher degree.

Take now an abelian integral on H_z

$$u = \int R(x,y,z)\,dx, \tag{1.2}$$

where $R \in C(F)$, the function field of F, is holomorphic at 0, and the path of integration is in H_z.

(1.3) Picard's problem. To find the behavior of the periods of u as functions of z in the vicinity of $z = 0$.

The solution to be given is essentially that of Picard [2], save that his recourse to the classical Picard-Fuchs equation has been replaced by an extensive application of the Weierstrass preparation theorem plus topology. This is justified on the following basis. The general Picard method is not only valid for absolute cycles as paths of integration but also for paths which are relative cycles with respect to boundaries "remote" from the singularities and this could only be done by utilizing topology of the neighborhood of the singularity. This will characterize all our later deviations from the method of Picard.

2. Method

In substance it will consist in replacing the collection of Riemann surfaces of the set H_z near the origin by a suitable Riemann representation of the neighborhood with the origin excluded.

Remark. It is clear that a ruled surface $F = 0$ offers no interest as regards to the Picard problem so that this special case is entirely ruled out in what follows.

(2.1) At the outset it is desirable to limit the discourse to analytic functions in a small neighborhood of the origin say

$$U:\ x\bar{x} + y\bar{y} + z\bar{z} < R^2.$$

Consider now $F = 0$ as an equation in y. The preparation theorem allows us to replace the equation of the surface in a suitable

U by a special polynomial in y

$$w = y^2 - 2\phi(x,z)y + \psi(x,z) = 0$$

where

$$EF \equiv w.$$

By identifying one obtains

$$\phi = -b'z + \cdots, \quad \psi = z + x^2 + 2a'xz + c'z^2 + \cdots .$$

Hence in U one has the two solutions:

$$-b'z + \cdots \pm(b'^2z^2 + \cdots - (z+x^2+2a'xz+c'z^2+\cdots))^{1/2}. \tag{2.2}$$

Let $y_1(x,z)$, $y_2(x,z)$ designate these two roots. They become equal when the radical is zero. Once more the preparation theorem shows that this occurs when

$$x^2 + 2(c''z+\cdots)x + z + \cdots = 0. \tag{2.3}$$

The two $x(z)$ roots of this equation are given by

$$\begin{aligned} x &= -c''z + \cdots \pm(-z+\cdots)^{1/2} \\ &= -c''z + \cdots \pm(-z)^{1/2}(1-1/2\ c^2z+\cdots) \\ &= z^{1/2}(\pm i - cz^{1/2} \mp \tfrac{i}{2}\ c^2z+\cdots). \end{aligned} \tag{2.4}$$

These two power series in $z^{1/2}$ have a common radius of convergence $\sqrt{\beta}$ (β real and > 0). Hence for $|z| < \beta$ they represent the branch points $x_1(z)$ and $x_2(z)$ of $y(x,z)$ as a function of x for z fixed, on the curve H_z. Of course both $\to 0$ with z.

At this point an important observation must be made. If a path of integration λ for the integral passes on H_z <u>between</u> the two branch points, it will not be possible to consider the behavior of

$$\int_\lambda R dx$$

as z goes through zero. The path will have been <u>pinched</u> as the physicists have it, by the point of contact of $z = 0$ with the surface $F = 0$. This is also called a <u>pinch point</u>: it is its first appearance in the literature. Since one may not continue λ through $z = 0$, Picard adopted the standard procedure since he utilized it: replacing the passage through $z = 0$ by the description of a small circle around $z = 0$ in the z plane. This procedure will be applied in a moment.

Returning to our two branch points $x_1(z)$ and $x_2(z)$, there is manifestly an $\alpha > 0$ such that for $|z| \in (0,\alpha)$ the two branch points remain distinct. Our first step is to impose $2R < \alpha,\beta$ that is $2R < \inf(\alpha,\beta)$. As a consequence on all the curves H_z to be considered the branch points $x_1(z)$ and $x_2(z)$ will be distinct.

There is no interest in viewing the intersection $H_z \cap U$ for $|z| = 2R$, since then in U: $x = y = 0$. We shall limit, therefore, $|z|$ to the interval ℓ: $0 < |z| < R$ ($|z| \in (0,R)$). Choose a fixed z such that $|z| = \rho \in \ell$. We wish to construct explicitly the "quasi-Riemann surface" Φ_z for $H_z \cap U$. To that end we follow the usual "lacet-construction" limited to this part of H_z.

3. <u>Construction of the Lacets of the Surface</u> Φ_z

In the x plane select the starting point A: $x = 0$ for the lacets. Under our assumptions the point $A \neq x_1(z), x_2(z)$: these two branch points are distinct. Since they permute the same values $y_1(x,z)$ and $y_2(x,z)$ the Riemann surface Φ_z will consist of two lacets $L_1 = Ax_1$ and $L_2 = Ax_2$, so far not directed. Take a circle C_ρ in the x plane centered at A and whose radius exceeds the largest distance from A to the x_i for every $|z| = \rho$, by an amount ρ. Thus $C_\rho \to 0$ as $z \to 0$.

Let now $y_1(A)$, $y_2(A)$ be the values of $y(x)$ at the point A.

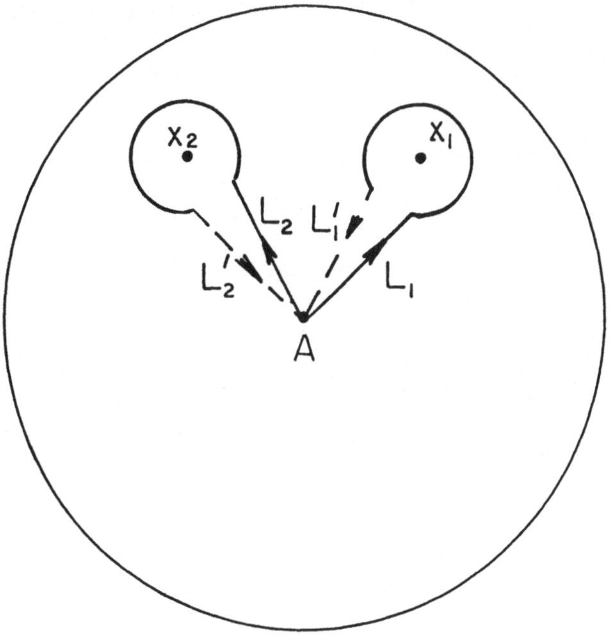

Figure 1.

The values are uniquely determined for the whole of the interior of C_ρ minus the lacets L_i. Continue to denote by L_i the lacet L_i followed with the value y_1 from A_1 to x_i and returning after a positive rotation around x_i back to A_2 (dotted line in Figure 1.) with the value y_2. Let also L_i' denote the lacet L_i starting from A_2 with y_2 (dotted line in Figure 1.) and after a positive rotation around x_i returning to A_1 with y_1. The surface Φ_z is obtained in the following manner. Let $C_{\rho i}$ denote the image of C_ρ in the sheet $x_i(z)$ corresponding to $y_i(x,z)$ $(|z| = \rho)$ and let $\Omega_{\rho i}$ denote the image of $|x| \leq \rho$ in the sheet $y_i(x,z)$ $(|z| = \rho)$. The surface Φ_z is now constructed by copying the ordinary construction of a Riemann surface limited however to the closed regions $\Omega_{\rho i}$, thus it is bounded by the circles $C_{\rho i}$.

4. Cyles of Φ_z. Variations of Integrals Taken on Φ_z

It is now evident that there exists in Φ_z a one-cycle δ^1 which $\to 0$ with z. Moreover, as z describes positively the circle D_ρ: $|z| = \rho$ in its plane the positive lacet L_1 augments by exactly the one-cycle δ^1.

In other words we have

$$\text{Var } L_1 = \delta^1 = (\delta^1, L_1)\delta^1. \tag{4.1}$$

Similarly

$$\text{Var } L_2 = \delta^1 = (\delta^1, L_2)\delta^1.$$

Hence if c is any pinched chain we will have

$$\text{Var } c = (\delta^1, c)\delta^1. \tag{4.2}$$

This is the basic formula that we had in view. It expresses the Picard-Lefschetz theorem.

As regards $\int_c$ its variations then defined by

$$\text{Var} \int_c = (\delta^1, c) \text{ per } \delta^1.$$

5. An Alternate Proof of the Picard-Lefschetz Theorem

The proof just given for the theorem is adequate for the theorem proper but not for others to follow. For the present, we confine our attention to an isolated singularity for convenience located at the origin.

Since F is a nonunit by the theorem of Weierstrass we may write

$$EF \equiv \prod F_h^{s_h} \tag{5.1}$$

where the F_h are locally irreducible and relatively prime special polynomials in y. Moreover since H_z for z small and $\neq 0$

has no singular points we may select R so small that in U any two distinct factors F_h, F_k are only jointly zero at the origin. Hence as far as the behavior in U is concerned we may merely consider one of the factors F_h. That is, we may merely discuss

$$F \equiv y^p + \alpha_1(x,z)y^{p-1} + \cdots + \alpha_p(x,z) = 0 \tag{5.2}$$

where the right hand side is an irreducible special polynomial in y. Our first task will be to reconstitute the topological features of M.

Evidently $F \equiv y^{p-1}(y+a_1)$ is without interest. Hence the discriminant of F as to y is only of interest when it has a special polynomial factor in x of positive degree. Let the product of its simple roots in x (evidently a special polynomial in x) be

$$\Delta(x,z) \equiv x^q + \beta_1(z)x^{q-1} + \cdots + \beta_q(z). \tag{5.3}$$

The equation $\Delta = 0$ represents the distinct <u>branch curves</u> of the solutions of (5.2). What interests us, however, is primarily the branch points of the solutions of (5.2) for z fixed and small. The general Puiseux analysis is applicable here. We infer from it that for $|z|$ small enough these solutions are given by a finite set of fractional power series in z. Let one of these be

$$x = z^{\lambda/r}(\gamma_o+\gamma_1 z^{1/r}+\cdots) = z^{\lambda/r}E(z^{1/r}), \quad \gamma_o \neq 0. \tag{5.4}$$

In fact, if we set $z = u^r$, and consider $z^{1/r}$ as any one of the solutions in u of this equation, x will represent a set of r branch points of (5.2), which are circularly permuted as z describes once the positive circle D_ρ of the z-plane given by $|z| = \rho$. We shall limit ρ in a moment. Let the successive branch points just considered be denoted by $a_h(z)$, $0 < h \leq r$.

We propose to examine the variation of the solutions of (5.2) near these branch points as z describes D_ρ. To that end set

$$x = \xi z^{\lambda/r},$$

$$\xi = \gamma_o + u + \gamma_1 z^{1/r} + \cdots, \qquad (|u| \text{ small}).$$

Upon substituting in (5.2) it assumes the form

$$y^p + \alpha_1(\xi z^{\lambda/r}, z)y^{p-1} + \cdots + \alpha_p(\xi z^{\lambda/r}, z) = 0.$$

This is a special polynomial in y with coefficients nonunits in $z^{1/r}$. Since each branch point only permutes two roots, the effect of z describing D_ρ with ξ fixed is to permute in each circular system the branch points circularly each with its two neighborhoods with order preserved.

Since for $|z|$ small and $\neq 0$ $x(z) = 0$ is not a solution of (5.4) we have certainly for all branch points

$$|x(z)| < \alpha|z^\nu|, \ \alpha > 0, \ \nu \text{ rational}.$$

However, the requirement $|x| < 2R$ imposes $2R > \alpha|z^\nu|$ hence

$$|z| \leq \inf\{R, (\tfrac{2R}{\alpha})^{1/\nu}\} = \zeta.$$

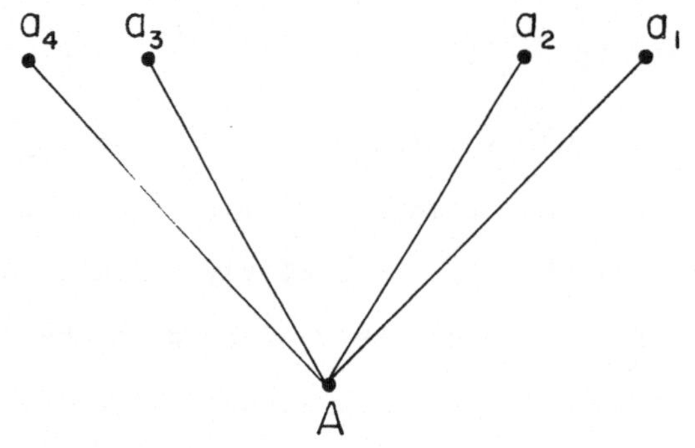

Figure 2.

On the complex x sphere S_x draw a positive circle C_ρ centered at the point A: $x = 0$ and of radius $\alpha\rho^\nu$, for $\rho \in (0,\zeta]$. For $|z| \in (0,\rho]$ let $a_h(z)$, $0 < h \le q$, be the critical points in S_x (all solutions of (5.4)). In S_x draw the lacets Aa_h. Let Ω be the open region between the lacets and C_ρ. Thus $\bar{\Omega} = \mathrm{St}\, C_\rho \cup \text{lacets}$ In Ω the values of the roots are uniquely determined by their values on C_ρ. Hence one may imitate the construction of the standard Riemann surface for the portion of H_z coresponding to $x \in \Omega$. The surface Φ_z may differ from a standard surface in that it has boundaries. Call their union Γ. We agree that the designation Φ_z represents the <u>open</u> surface. Thus $\bar{\Phi}_z = \Phi_z \cup \Gamma$.

We must also show that Φ_z <u>is</u> connected. In fact, in the contrary case let $y_h(x,z)$, $0 < h \le s < p$ be the collection of the related roots to one of its components. The symmetric functions of these roots are manifestly nonunits. Hence they are the roots of a special polynomial factor in y which is a factor of F of (5.1) – contrary to the assumption that F is irreducible. Thus Φ_z corresponds to F itself and is connected.

In $\bar{\Phi}_z$ the lacets, Ω and C_ρ will be repeated r times for y_h and denoted by Ω_h for Ω and $C_{\rho h}$ for C_ρ. However, in Ω_h one may omit the lacets corresponding to any branch points which do not permute y_h. Let now Γ designate the collection $C_{\rho h}$.

We note these properties:

(5.5) (a) Φ_z <u>is a connected complex mod</u> Γ <u>which is an orientable analytic</u> 2-<u>manifold (surface)</u> M^2. (b) <u>Each</u> Ω_h <u>is homeomorphic to a closed plane ring surface</u>.

Let G be the graph composed of all the lacets. From (5.5) we infer:

(5.6) <u>The graph</u> G <u>is a deformation retract of</u> $\bar{\Phi}_z$ <u>hence also of</u> Φ_z.

Agree to denote the one-cycles of Φ_z by δ^1 with eventual subscripts. Then (5.6) imples:

(5.7) *Every* $\delta^1 \sim$ *in* Φ_z *to a cycle* $\overline{\delta}^1$ *of* G. *If* $\delta^1 \sim 0$ *in* Φ_z *then* $\overline{\delta}^1 \sim 0$ *in* G *and hence* (*since* dim G = 1) $\overline{\delta}^1 \equiv 0$.

For a later purpose we need an extension of the preceding properties. Let A_h be the image of $x = 0$ (the point A of S_x) and draw an arc A_hB_h in Ω from A_h to a point B_h of $C_{\rho h}$. Let $G_1 = G \cup A_hB_h$.

(5.8) $\Omega_h - A_hB_h$ *is a* 2-*cell* E_h. *The set of these* 2-*cells plus their boundaries constitute a decomposition of* $\overline{\Phi}_z$ *into a cellular complex whose derived is simplicial*.

(5.9) *The graph* G_1 *is a deformation retract of* $\overline{\Phi}_z$.

Since G_1 consists of G plus the lacets confined to Γ along the arcs A_hB_h, we infer from (5.9):

(5.10) *Every cycle* γ^1 *of* $\overline{\Phi}_z$ mod Γ *is* $\sim$ *in* $\overline{\Phi}_z$ *to a cycle of* G_1 mod Γ.

We point out finally these properties of $\overline{\Phi}_z$ as an M^2 bounded by Γ:

(5.11) *Between the absolute cycles* δ^1 *of* Φ_z *and those* γ^1 *of* $\overline{\Phi}_z$ mod Γ *there exist the Betti number relation*:

$$R^1(\Phi_z) = R^1(\overline{\Phi}_z,\Gamma) = r. \tag{5.12}$$

(5.13) *From* (5.12) *one infers that one may choose bases* $\{\gamma_h^1 |\ 0 < h \leq s\}$, $s \leq r$, *and* $\{\delta_k^1 |\ 0 < k \leq s\}$ *for their types such that their intersection number matrix* $[(\gamma_h^1, \delta_k^1)]$ *is nonsingular*.

(5.14) We must now consider the variations of the cycles. This must be based upon an examination of the variations of the lacets. We are fixing our attention on the behavior of the surface Φ_z, where z starting from a position z_o on D_ρ describes this circle once.

Let Aa denote a general lacet and suppose that a permutes y_h, with y_j as one turns positively around a. We temporarily denote the path Aa from A to a by (h,j). If such a path is part of a closed path it will necessarily be followed by (j,k).

Suppose now that as z describes D_ρ as described above, a_1 goes into a_4. Upon deforming a_1 in the positive direction around the branch points it will assume a new position say Aa_4 (Figure 3)

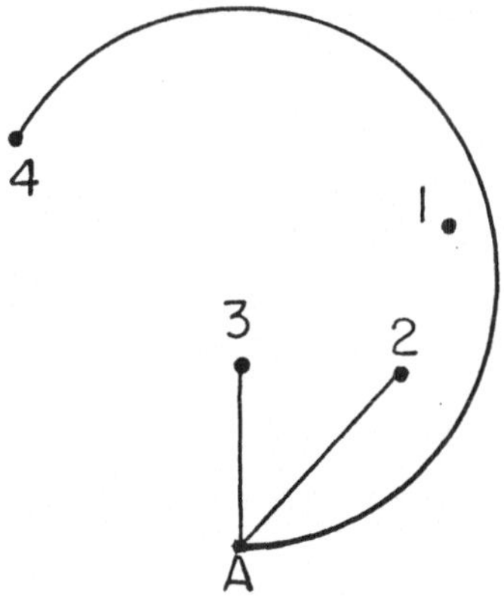

Figure 3.

This new lacet may be deformed into a new set as shown in Figure 4.

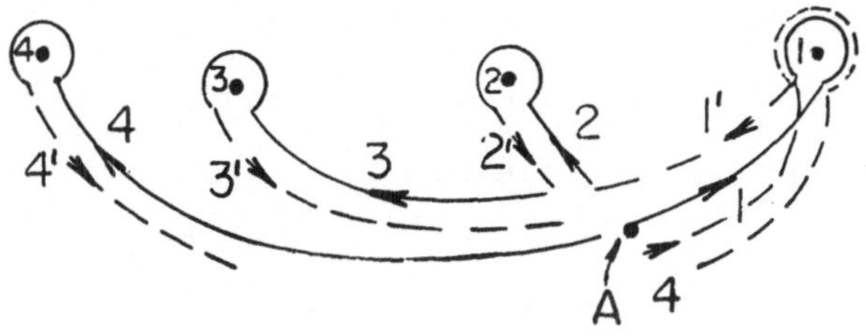

Figure 4.

This may be represented schematically by the expression

$$(1,2') + (2,3') + (3,4') + (4,3') + (3,2') + (2,1') + (1,2').$$

Hence we have:

$$\mathrm{Var}(1,2') = (1,2') + (2,3') + \cdots + (2,1') = \delta_j^1$$

which is a one-cycle of Φ_z. In the same manner we have

$$\mathrm{Var}\, Aa_j = \delta_j^1.$$

In the same manner as in section 4, we show that

$$\mathrm{Var}(Aa_j) = (\delta_j^1, Aa_j)\delta_j^1$$

and finally for any cycle γ^1 of Φ_z mod G_1:

$$\mathrm{Var}\,\gamma^1 = \Sigma(\delta_j^1, \gamma^1)\delta_j^1. \tag{5.15}$$

This is almost our basic variation formula. We say "almost" because Φ_z is really just a component of an earlier Φ_z. It is immediately apparent, however, that the same expression (5.15) is adequate.

6. The Λ_1-manifold M. Its Cycles and Their Relation to Variations

In a sense the Φ_z's just considered are only indirectly related to the singular point at the origin. It is of interest, therefore, to replace them by a more closely related scheme.

When z describes the circle D_ρ each $C_{\rho h}$ generates a torus T_h and hence Γ generates a finite set of such tori. As $|z|$ describes the segment $[0,\zeta]$ each torus T_h generates a cone-like variety with vertex at the origin and base T_h for $|z| = \zeta$. The union of these varieties is Λ_1.

The union of the tori of the T_h for $|z| = \zeta$ and the origin is a set denoted by Λ_2.

When $|z|$ describes the interval $(0,\zeta)$, $\Phi_z - \Gamma$ generates an orientable analytic 4-manifold M. Clearly

$$\partial M = \Lambda_1 \cup \Lambda_2.$$

One may cover $\bar{M}$ with a subcomplex of F, still written $\bar{M}$, with F so chosen that Λ_1, Λ_2 are normal in $\bar{M}$. From the fundamental duality theorem (I.12.8) we have then

$$R^1(M-\Lambda_2,\Lambda_1) = R^3(M-\Lambda_1,\Lambda_2). \tag{6.1}$$

(6.2) <u>Lemma</u>. <u>Let</u> δ^3 <u>denote any</u> 3-<u>cycle of</u> $M - \Lambda_1 \bmod \Lambda_2$ <u>and let</u> $\delta^1 = \delta^3 \cdot \Phi_z$. <u>Then there is equivalence between the homologies</u> (a) $\delta^1 \sim 0$ <u>in</u> $\Phi_z - \Gamma$; (b) $\delta^3 \sim 0$ <u>in</u> $M - \Lambda_1 \bmod \Lambda_2$.

The second relation obviously implies the first. There remains to prove (b).

From (5.7) there follows that we may assume that δ^1 is a cycle of the graph G. As such when z varies δ^1 generates a certain $\bar{\delta}^3$. Evidently if δ^1 is a cycle of $\bar{\Phi}_z \bmod \Gamma$ then

$$(\gamma^1,\delta^3)_M = (\gamma^1,\delta^1)_{\Phi_z}.$$

Hence

$$(\gamma^1,\delta^3-\bar{\delta}^3) = 0; \quad (\gamma^1,\delta^3)_M = (\gamma^1,\bar{\delta}^3)_M.$$

Now $\delta^1 \sim 0$ in Φ_z implies $\delta^1 \equiv 0$, hence $\bar{\delta}^3 \equiv 0$ and so

$$(\gamma^1,\delta^3) = 0$$

whatever γ^1. Therefore $\delta^3 \sim 0$ in $M - \Lambda_1 \bmod \Lambda_2$. This proves the lemma.

(6.3) <u>Corollary</u>. <u>If</u> $\{\delta_h^3 \mid 0 < h \leq r\}$ <u>is a base for the</u> δ^3 <u>cycles then</u> $\{\delta_h^1 = \delta_h^3 \Phi_z \mid 0 < h \leq r\}$ <u>is a base for the one-cycles of</u> Φ_z. <u>Hence a base</u> $\{\gamma_h^1\}$ <u>for the cycles of</u> $\bar{\Phi}_z \bmod \Gamma$ <u>is likewise a</u>

base for the one-cycles of $M - \Lambda_2 \bmod \Lambda_1$.

(6.4) Hence every one-cycle γ^1 of $M - \Lambda_2 \bmod \Lambda_1$ depends upon a one-cycle of $\Phi_z \bmod \Gamma$ (whatever Φ_z, $|z| \in (0,\zeta)$).

Finally, if δ_h^3 is the 3-cycle of $M - \Lambda_1 \bmod \Lambda_2$ such that $\Phi_z \delta_h^3 = \delta_h^1$ (δ_h^1 vanishing cycle) then:

$$\mathrm{Var}\ \gamma^1 \sim \sum (\delta_h^3, \gamma^1)\delta_h^1 \quad \text{in} \quad M - \Lambda_1 \bmod \Lambda_2.$$

This is the "M variation expression" which we had in view.

CHAPTER III

EXTENSION TO HIGHER VARIETIES

1. Preliminary Remarks

A problem entirely analogous to the extension of a one-dimensional integral beyond an isolated singularity arises for higher dimensional integrals. What is required is an analysis of the behavior of the k-cycles of the sections H_y of a V^{k+1} around the isolated singularity. (We write now y instead of the earlier $\underline{z}$.) As it happens the performance concerned with an irreducible algebraic variety U^r in the space $X \times Y$ ("irreducible and "algebraic") have their obvious meaning. However, the varieties $\underline{y}$ = const. will only play a minor role. To simplify matters we assume that the generic varieties (dimension p) have no singularities. Our purpose will be precisely to study the effect on certain algebraic integrals

$$J(\underline{y}) = \int_{\Gamma} R(\underline{x},\underline{y})\, d\underline{x}$$

of the eventual singularities in the x varieties H_y. Their singularities are characterized by a locus in the y space known as the <u>Landau variety</u> whose definition is fairly clear and need not be described explicitly. We designate it by the letter L. Historically, L has been introduced by the distinguished Moscow theoretical physicist Liov Davidovich Landau who died a couple of years ago, a victim of an automobile accident.

It is evident that L is an algebraic variety of the space Y and we will continue to take full advantage of this fact.

<u>We state explicitly that the points of</u> Y - L <u>correspond to the nonsingular varieties</u> H_y, <u>while those of</u> L <u>correspond to the</u>

singular varieties H_y.

2. First Application

In the preceding part we have dealt extensively with the cases $p = q = 1$. The next most interesting case is $p = 1$ and $q > 1$. In other words it is the first case when the Landau variety is not of dimension zero. The varieties H_y continue to be algebraic curves (nonsingular or general).

Let Y_* denote a simplicial covering complex of the space Y. Since L is algebraic one may cover it also by a simplicial normal subcomplex L_* of a suitable subcomplex of Y_*. Take now $\mathrm{St}\, L_*$ (in Y_*). If P is any point of $\mathrm{St}\, L_* - L_*$ we have seen that there is a unique segment QR with $Q \in L_*$ and R $\overline{\mathrm{St}\, L_*} - \mathrm{St}\, L_*$.

Let the ratio $PQ/PR = k \geq 0$. The locus of all the points P for a fixed k, $0 < k < 1$, is the S_k^c of L. It is evidently a polyhedral complex with the following properties:

(a) All the S_k^c, are homeomorphic.

(b) Since Y is an M^{2q} all the S_k^c are absolute orientable M^{2q-1}.

(c) S_k^c is a deformation retract of both L and the locus of the point R.

(d) Let S_k, $k_1 \leq k$, $= \{S_h^c |\ 0 \leq h \leq k\}$. Since $q \geq 2$, $\dim S_k^c \geq 3$. Hence one may construct the Poincaré group $\pi(Y - L - S_k^o)$ issued from a fixed point A_k of S_k^c so that no two of its generators (finite in number) meet at any other point than A_k.

(2.1) Let now X_* denote a simplicial complex covering X and whose mesh $< \varepsilon$ (edges of length $< \varepsilon$). Let $\mathcal{S}$ denote the set of singular points of the curves H_y (they correspond to the points of the Landau variety L in Y). Since $\mathcal{S}$ is algebraic it has a covering subcomplex $\mathcal{S}_*$ of X_* and we may assume that it is a normal subcomplex of X_*.

(2.2) Let now g_k be an element of the Poincaré group of $Y_* - \text{St } L$ issued from A_k. It is now evident that there exists a close parallel between the arguments of II, Sections 2 and 3.. to the end of Chapter II. It is only necessary to point out the differences:

(a) the operation g_k takes the place of C_ρ of Section 1 and on;

(b) the number k takes the place of ρ;

(c) as $k \to 0$ the vanishing one-cycles δ^1 are replaced by the one-cycles of $\mathcal{S}$;

(d) of course $\mathcal{S}$ takes the place of the isolated singularity of the surface F;

(e) h-cycles mod Λ_i are replaced by analogous cycles easily defined;

(f) the variation formulas are also modified in an obvious way.

3. Extension to Multiple Integrals

It is clear that what has been done so far for simple integrals on H_y also may be dealt with for multiple integrals on H_y. However, there arise new considerations which must be discussed separately. It so happens that the key to the real questions will already occur for double integrals and so we discuss them first.

(3.1) Topologically the new problem is actually how to continue a real double field of integration across a complex isolated singularity in the complex X-space. It is, therefore, necessary to study the V^3 case of X complex somewhat fully. This requires, however, the study of the behavior of the 2-cycles of a surface around a point of contact of an ordinary tangent plane of a surface in $\mathcal{P}^3$.

This as it happens has already been done by Picard. All that is necessary, therefore, is to give a reasonably extensive résumé of his results.

4. The 2-Cycles of an Algebraic Surface

For simplicity identify the surface F with the earlier surface $F(x,y,z) = 0$, $F(0,0,0) = 0$. We call again H_z the sections $z = \text{const.}$ and assume once more that for $|x|$, $|z|$ both small enough there are ν distinct values, $y_h(x,z)$, $1 \leq h \leq$, such that the planes $y = y_h$ are tangent to F, with two distinct tangents to H_z.

Now in the plane of x draw ν lacets from a fixed point A to the points x_h, A_h, and let δ_h^1 be the vanishing cycle of H_z from the contact B_h of $z = z_h$ with F. Assume that $|B_1 - B_2| \to 0$ with z. This implies in H_{az} that the cycles $\delta_1^1 \sim \delta_2^1 \bmod H_\infty$. Call δ_i^2, the chains generated by y as it describes AB_i, $i = 1,2$. Denote by H_A the section $y = A$. It is evident that $\delta_1^1 \sim \delta_2^1$ in $[H_A]$ (finite). Hence $\delta^2 = \Delta_1^2 - \Delta_2^2 - [H_A] \sim 0$ in H_A. Therefore, γ^2 is a vanishing cycle of H_z as $z \to 0$. Hence when z turns positively around zero in its plane.

$\text{Var}\ \Delta_1^2 = \delta^2 = -\text{Var}\ \Delta_2^2$. From this follows readily that if c^2 is a 2-chain in H_z, then under the same conditions

$$\text{Var}\ c^2 = -(\delta^2, c^2)\delta^2.$$

Except for the sign this is the same result as already obtained for one-chains.

More generally for V^{n-1}, $n \geq 2$, one will obtain

$$\text{Var}\ c^n = (-1)^{n-1}(\delta^n, c^n)\delta^n.$$

(all taken in any H_z).

Hence finally for any n-chain near the singular variety $\mathcal{S}$,

$$\operatorname{Var} c^n = \sum_{j=1}^{\nu} (-1)^{n-1} (\delta_j^n, c^n)\, \delta_j^n \tag{4.1}$$

where the δ^n tend to a specific n-cycle of $\mathcal{S}$. It is really not necessary to go beyond.

CHAPTER IV
FEYNMAN INTEGRALS

1. On Graphs

The structure of a Feynman integral is dictated by a certain special graph. I must, therefore, discuss these graphs.

A graph G is merely a real one-complex. It consists then of nodes $n_1, n_2, \ldots, n_{\alpha_0}$ and branches $b_1, b_2, \ldots, b_{\alpha_1}$. The general assumptions are:

I. G is finite, connected and inseparable: the removal of any node does not disconnect G.

II. Every branch has two distinct end nodes. The number of branches attached to any node n_j is the order of n_j. It is always ≥ 2.

III. Every branch is oriented, that is one fixes its initial node n' and terminal node n''.

(1.1) Arcs, loops, trees. An arc of G is a succession of distinct nodes and branches

$$n_1 b_1 n_2 b_2 \cdots n_k b_k n_{k+1}$$

where b_h joins n_h to n_{h+1}.

Note that "G is connected" means that any two of its nodes may be connected by an arc.

A loop is like an arc except that $n_1 = n_{k+1}$.

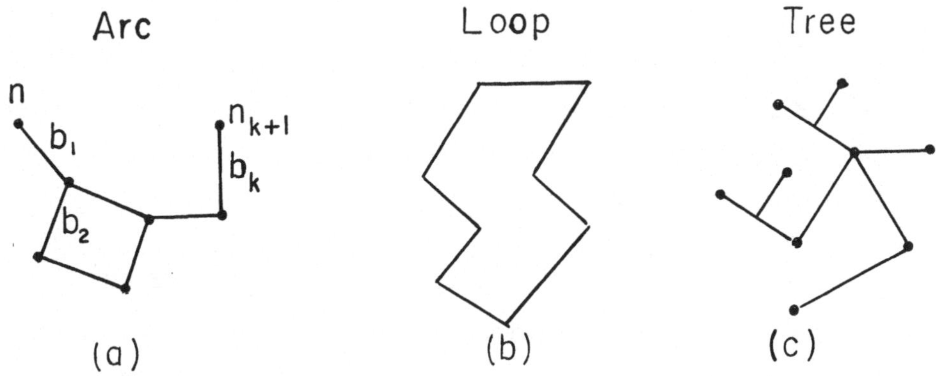

Figure 5.

A tree is a (connected) graph without loops. A <u>forest</u> of G is a finite set of trees of G. A <u>maximal tree</u> or <u>forest</u> T of G is one which ceases to be one when augmented by any branch of G.

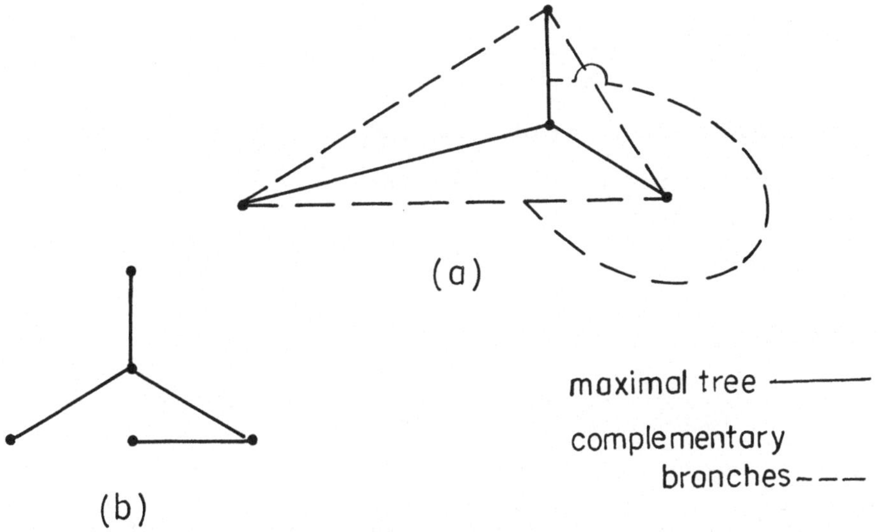

Figure 6.

(1.2) A maximal tree of G contains all the nodes of G. (Proof elementary.)

The boundary of a triangle is a loop. The broken line between two vertices is an arc.

(1.3) Remarks. A maximal tree of a graph G need not be unique. Thus if G is a polygon the removal of any side leaves a maximal tree, and is certainly not unique.

(1.4) Connectedness of G implies that the choice of T as maximal tree is unique. For suppose that another tree T' of G is also maximal. By (1.2), and since G is connected T and T' have the same nodes. Since $T' \neq T$ some pair (n,n') of nodes of G must be common end nodes for $b \in T$ and $b' \in T'$, $b \neq b'$. But this implies that $b \cup b'$ is a loop of $T \cup b'$, and therefore, that T is not maximal. Therefore, T' does not exist and T is unique.

2. Algebraic Properties

They are, a combination of orientation and a basic abelian group of coefficients $\mathcal{G}$: frequently the additive group of rational numbers, most usual, and later that of quaternions.

Two linear collections of nodes and of branches will play an important role. They are the zero-chains

$$c_o = \sum x_j n_j$$

and one-chains

$$c_1 = \sum y_k b_k$$

x_j and $y_k \in \mathcal{G}$.

A branch b of G has assigned an initial node n' and a terminal node n". A definite selection is described by the symbol $\vec{b}$

and its opposite may be denoted by $\overleftarrow{b}$ or $-\vec{b}$. Define $n'' - n'$ as the boundary of $\vec{b}$ written

$$\partial\vec{b} = \partial(-(-\vec{b})) = n'' - n'.$$

Hence by linear extension

$$\partial c_1 = \sum y_k \vec{b}_k = d_j n_j.$$

Fixing the arrow in this conventional way is called orienting b.

Whenever $\partial c_1 \equiv 0$, that is every $d_j = 0$, one refers to c_1 as a one-cycle over G, or merely a cycle (understood with $\mathcal{G}$ the rationals). The maximum number R of linearly independent one-cycles (usually over the rational numbers) is the first Betti number of G. This is a very important number in all that follows.

(There is an analogue Betti number R_o, for zero cycles (= zero-chains) but in a connected graph $R_o = 1$.)

(2.1) Characteristic. It is defined as

$$\chi(G) = \alpha_o - \alpha_1.$$

$\chi(G) = 1 - R$ (Poincaré).

Let $\vec{b}_j$ be a branch and n_k a node. Define an incidence number η_{jk} as +1,-1 or zero accordingly as n_k is terminal, initial, or not one of the two types. One refers to $[\eta]$ as incidence matrix of G.

(2.2) Kirchoff's current law (electric current). For every node n_k in a cycle $\gamma = \Sigma\, y_j b_j$

$$\sum y_j n_{jk} = 0.$$

That is: the numbers of arriving and departing currents to any node are equal.

For this reason a one-chain is sometimes referred to as current.

(2.3) Every cycle γ contains a loop.

That is with some of the elements of γ one may arrange a loop.

Let $\gamma = \Sigma y_j b_j$. Suppose that $y_1 \neq 0$ so that b_1 is a branch of γ and let n'' be its terminal node. From Kirchoff's law there follows that there is a b_2 in γ with the initial node n''. Similarly beyond b_2, etc. That is there is obtained a succession of elements of

$$n_1 b_1 n_2 b_2 \cdots n_k b_k n_{k+1}$$

forming an arc $\subset \gamma$. However, since the construction must terminate for some k: $n_{k+1} = n_1$, resulting in a loop in γ.

(2.4) Observe in passing that the arc above may have its branches oriented so that

$$\partial \vec{b}_h = n_{h+1} - n_h$$

which produces for the total arc α the orientation

$$\partial \vec{\alpha} = n_{k+1} - n_1.$$

The result for the loop λ is an orientation $\vec{\lambda}$ turning it into a cycle. Then reversing all the orientations produces the cycle $-\vec{\lambda}$.

(2.5) With the same G,T, the difference $G - T$ consists of a set of distinct branches $\{b_h^* \mid 1 \leq h \leq \rho\}$ with their end nodes (n',n'') in T. Since T is connected n' may be joined by an arc τ_h of T to the node n''. Let $\vec{b}_h$ and $\vec{\tau}_h$ be oriented from n' to n'' for $\vec{b}_h$ and from n'' to n' for $\vec{\tau}_h$. Hence

$$\vec{\lambda}_j = \vec{b}_h^* + \vec{\tau}_h$$

is an oriented loop of G. _We refer to_ $\vec{\tau}_h$ _as a junction of_ $\vec{b}_h^*$. _There is exactly one junction of each_ $\vec{b}_h^*$.

For if there were another say $\vec{\tau}_h'$, $\vec{\tau}_h - \vec{\tau}_h'$ would be a loop of T.

(2.7) $\{\vec{\lambda}_h\}$ _is an additive base for the cycles of_ G. _Hence the number of distinct_ $\vec{\lambda}_h$ _is_ R _for all trees_. _In particular_, $\rho = R$.

This will follow from these two properties:

(a) _Every cycle_ γ _is a linear combination of the_ $\vec{\lambda}_h$.

(b) _The cycles_ $\vec{\lambda}_h$ _are independent_.

Proof of (a). Since γ is a cycle it cannot be in T. Thus

$$\gamma = \sum u_h \vec{b}_h^* + v, \qquad v \subset T, \text{ the } u_h \text{ not all zero.}$$

Since

$$\vec{\lambda}_h = \vec{b}_h^* + w, \qquad w \subset T$$

we have

$$\gamma = \sum u_h \vec{\lambda}_h + \mu, \qquad \mu \subset T.$$

However (by difference) μ is a cycle and hence $\mu = 0$. This proves (a).

Proof of (b). A relation

$$\sum y_h \vec{\lambda}_h = 0$$

with the y_h not all zero implies that, in fact

$$\sum y_h \vec{b}_h^* = 0$$

which is rules out since the $\vec{b}_h^*$ are all distinct. Hence (b) holds.

(2.8) Let C, Z denote the additive groups of chains and cycles over integers or any field. Since $Z \subset C$, there is a factor group $C/Z = C_o$, the group of pure chains: none of its non-zero elements is a cycle. Thus

$$C = Z \oplus C_o. \tag{2.9}$$

(2.10) Any two distinct loops λ, λ' have at most in common a single arc or a single node.

For the intersection can only be an arc, not a b_h^*, or a node, and in both cases part of T. However, more than one intersection element chain or node would imply that T has a loop.

3. Feynman Graphs

The study of Feynman integrals calls for two important complements: a new type of graph, and a considerable broadening of the type of allowed coefficients.

(3.1) We first deal with the new graphs. Corresponding to our previous graph G there is an ampler graph H, with G as a subgraph. H has one more node n_o than G and new branches $\{b_{ej}\}$, $0 \le j \le N$,

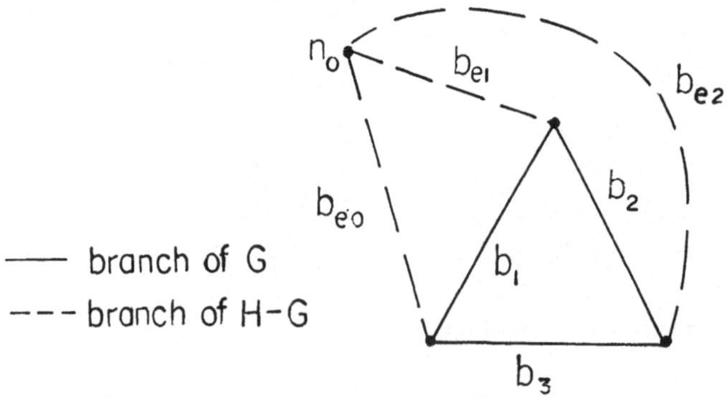

Figure 7.

joining n_o to every node of G. The b_{ej} are oriented from n_o out. n_o and the b_{ej} are exterior elements and those of G are interior elements.

(3.2) Kirchoff's law holds for all H and notably at n_o. Hence the only relation between the currents $\vec{b}_{ej}$ is

$$\sum \vec{b}_{ej} = 0. \tag{3.3}$$

The currents β_j in the b_{ej} are not meant to contribute to the formation of new loops but rather in supplementing the currents in the internal branches. This is illustrated by Figure 8.

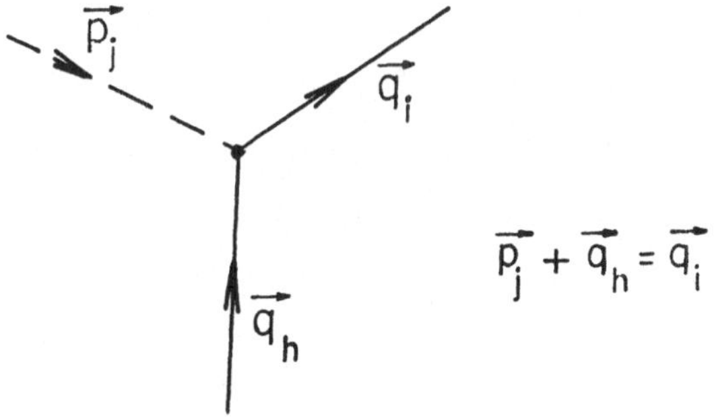

Figure 8.

(3.4) In the Feynman theory there come up quaternions as coefficient group. This gives rise to new chains and cycles and related results will continue to hold. The only one requiring some argument is (2.3): every cycle γ contains a loop. Now a quaternionic cycle γ has actually four components behaving like the cycle itself. In particular, the first component, say γ_1, has exactly the same structure as the associated "ordinary" cycle. Hence γ_1 has a loop and so does the quaternionic cycle itself.

As is well known quaternions lend themselves naturally to defining zero, addition and subtraction. Difficulties occur however with multiplication and division. Multiplication requires this deviation. Suppose that we deal with a collection $\{\gamma_h | 1 \leq h \leq r\}$ of quaternion cycles say relative to $\mathcal{G}$:

$$\gamma_h = \sum x_{hj} q_j .$$

The γ_h are dependent or independent accordingly as there exist rational elements η_h such that

$$\sum_{h,j} \eta_h x_{hj} q_j \equiv 0$$

$$j = 1,2,3,4.$$

4. Feynman Integrals

They appear as the coefficients in a power series which occurs in particle physics. Their number is, therefore, infinite but we shall only treat a single integral.

The general form is:

$$F(\underline{p},\underline{m}) = \int_{\Gamma^{4R}} \frac{\prod_h d^4 k_h}{\prod_i Q_i} . \tag{4.1}$$

See, however, Section 9.

Here F is an analytic function of parameters p_j and m_h. The m_h are real masses and generally distinct.

The integral is taken over a real 4R-space. The integration may have to omit certain singularities by passing to an appropriate complex and merely local space. (For details see Section 10.)

The notation

$$d^4 k_h = dk_{h1} dk_{h2} dk_{h3} dk_{h4}$$

(quaternionic components of k).

Q_h is a quadratic polynomial in the components of q_h (branch b_h of G). Similarly p_i corresponds to the exterior branch b_{ei}. The exact expression is

$$Q_i = |q_i^2| - m_i^2$$

(quaternionic distance).

Restriction. Any two loops λ, λ' intersect in at most one node or one arc.

(4.2) Currents in the branches b_h^*. Actually this current $q_h^* = k_h$. Hence

$$\frac{\partial h_h}{\partial q_h^*} = 1,$$

an important relation in a moment.

5. Singularities

The feature that will attract our attention is the variation of the function F in a neighborhood of a singularity. We first discuss the problem for a rather simple situation.

(5.1) The first observation is that the range R in the differentials leaves one free to modify the order of the dk_h. We assume that as the first move the first R Q_h are in the same order as the dk_h, that these Q_h for adjustable parameters m_h are all distinct, and the same Q_h occur only once. Moreover, we assume that in (4.1) and at least as regards our "singularity problem" one may integrate successively with respect to $dk_1, dk_2, \ldots, dk_R$. That is one may substitute for (4.1)

$$I = \int \prod_{h=1}^{R} \left(\frac{d^4 k_h}{Q_h} \right), \qquad 1 \leq h \leq R. \tag{5.2}$$

(5.2a) <u>This implies that for the present we assume that every</u> $Q_{R+j} \equiv 0$.

Referring to (4.2) one may replace in the integral k_h by q_h^*. That is

$$I = \int \prod_{h=1}^{R} \left(\frac{d^4 q_h^*}{Q_h} \right), \tag{5.3}$$

the integral being taken in succession as to q_h^*.

Take now a typical q_h^* and Q_h and call them q and Q. Thus I will be formally replaced by the succession of integrals (for $1 \leq h \leq R$)

$$I^* = \int \frac{d^4 q}{Q} .$$

Finally replace q and components by t,x,y,z. Thus

$$Q = t^2 - x^2 - y^2 - z^2 - m^2$$

where t represents energy.

Note also that

$$Q = t^2 - r^2 - m^2, \quad r^2 = x^2 + y^2 + z^2.$$

6. Polar Loci

The first step is to examine the effect of the single polar locus $Q = 0$ on the integral I^*. In terms of the new coordinate this locus may be written

$$t^2 - r^2 = m^2. \tag{6.1}$$

One recognizes in it the 4-dimensional analogue of the 2-sheeted hyperboloid of 3-space. Therefore (6.1) is seen to consist likewise of two sheets Σ^m, Σ'^m corresponding to $t \geq m$ and $\leq -m$. Their joint appearance is that of the hyperboloid and is well illustrated by

Figure 9.

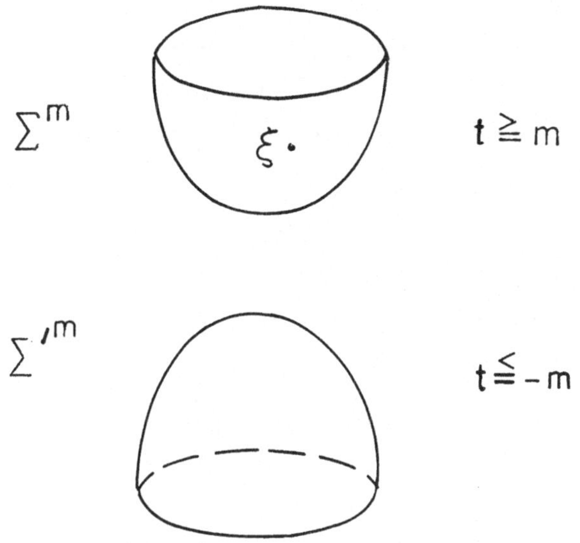

Figure 9.

For $m = 0, \Sigma^o$ and Σ'^o are the two sheets of a cone of revolution.

Take now any point $\xi(t_o, x_o, y_o, z_o)$ on Σ. We have to calculate the residues of $\frac{1}{Q(t,x_o,y_o,z_o)}$, $\frac{1}{Q(t_o,x,y_o,z_o)}$, etc. relative to the point ξ in the complex planes $\pi_{t_o}, \pi_{x_o}, \ldots$ parallel to the planes t alone complex, x alone complex, etc., through ξ.

The first residue is the integral $\int \frac{dt}{Q(t,x_o,y_o,z_o)}$ around ξ in π_t, that is

$$\int_{\pi_t} \frac{dt}{Q(t,x_o,y_o,z_o)}$$

around ξ in π_t. Now

$$Q(t,x_o,y_o,z_o) = t^2 - t_o^2,$$

$$\frac{1}{t^2-t_o^2} = \frac{1}{2t_o}\left(\frac{1}{t-t_o} - \frac{1}{t+t_o}\right).$$

Hence the required residue is $\frac{\pi i}{t_o}$. Similarly the other three residues are

$$\frac{-\pi i}{x_o}, \frac{-\pi i}{y_o}, \frac{-\pi i}{z_o}.$$

Hence the total residue of our integrals taken in succession is

$$-\frac{\pi^4}{t_o x_o y_o z_o} = \mathrm{Var}(t_o, x_o, y_o, z_o) \int \frac{d^4 q}{Q}.$$

The total variation of the whole integral is then

$$\mathrm{Var} \int \prod_{h=1}^{R} \left(\frac{d^4 q_h}{Q_h} \right) = \frac{(-1)^R \pi^{4R}}{\prod_h t_{oh} x_{oh} y_{oh} z_{oh}}.$$

(6.2) The residues and variations just calculated refer to residue and variations just around a point of the polar locus. The actual variation in the integral is to be obtained by integrating this local residue over the whole connected part of the polar locus.

The justification of this computation is as follows: It results from integrations in $\pi_t, \pi_x, \ldots,$ along small circles $\gamma_t, \gamma_x, \ldots,$ centered at $t_o, x_o, \ldots,$ in $\pi_t, \pi_x, \ldots$. Take points $\xi_t, \xi_x, \ldots,$ on the complex arcs of $\gamma_t, \gamma_x, \ldots,$ and join each to a fixed complex point ξ_o near ξ by arcs $\gamma_t, \gamma_x, \ldots$. The four closed paths

$$(-\vec{\delta}_t)\vec{\gamma}_t\vec{\delta}_t,\ (-\vec{\delta}_x)\vec{\gamma}_x\vec{\delta}_x, \ldots,$$

are basic paths of the Poincaré group of paths relative to ξ_o and the local complex extensions of the spaces $\pi_t, \pi_x, \ldots$. This justifies our calculation of the residues and variations.

(6.3) Polar locus $Q_j = 0,\ j \leq R,$ when some of the factors $Q_j,\ 1 \leq j \leq R$ coincide.

This may occur notably when some of the masses m_j are equal.

Assume that for a certain collection $\{Q_1,\ldots,Q_s\}$, $s \leq R$, the exponents are always unity, but that for all other cases they are > 1. We may then range the Q_j in such an order that for $1 \leq j \leq s$ the exponents, all distinct, are always unity, but for $j > s$ they are always > 1. Set

$$\Psi = \prod_{j>s} Q_j, \quad \Omega = \prod_{j\leq s} Q_j.$$

One may deal with Ω and its integration as before. One obtains residues for each factor, and none for the factors of Ψ. Further details are obvious and need not be discussed.

(6.4) Returning to the integral (5.3), in the ultimate integration its limits must be $-\infty,+\infty$. Recall that in this integration the complete polar locus $Q = 0$ of that integral must be avoided. This means that we must avoid the pair (Σ,Σ') by going to the complex domain. To that effect we take a sort of tubular neighborhood N_σ of the pair in their complex neighborhood, and replace integration through Σ and Σ' by complex integration along the boundary B_σ of N_σ.

(6.5) Return now to the complete integral, that is with properly indexed variables t_h,x_h,y_h,z_h, we merely have to replace (Σ^m,Σ'^m) by the cartesian product

$$P_h(\Sigma^{m_h},\Sigma'^{m_h})$$

together with

$$P_h B_{\sigma_h}$$

but without the neighborhoods N_{σ_h}.

(6.6) We now proceed to put back — in their place — the Q_j for

$j > h$. These Q_j consist of those corresponding to loop branches other than the basic branches q_h^* of the loops. In these branches (through loop intersections) there may appear diverse q_h^*. The all important fact, however, is the possible presence of elements p_j of the exterior branches. One of these branch currents already labelled may be described as $q_\ell^o + p_\ell^o$ where q^o may depend on several q_h^* and p_ℓ^o upon several initial p_i. The actual current is actually a parameter. We set then

$$q_\ell' = q_\ell^o + p_\ell^o, \quad Q_\ell' = [q_\ell'^2] - m'^2$$

the combination referring to parameters, and outside the basic integral.

To sum up then the full integral may now be described as

$$I = \frac{1}{\prod_\ell Q_\ell'} \int_{\Gamma^{4R}} \prod_{h=1}^{R} \left(\frac{d^4 q_h^*}{Q_h} \right)$$

where the part outside the integral consists solely of parameters.

7. More General Singularities

We shall now attack the same problem, but with a much more general point of view.

One must first ask what is a singularity? Since the question bears upon the performance of the Feynman integral (4.1) a singularity may be interpreted as referring to an integral for which there arises a new infinity with a polar residue, or rather one beyond the "normal" situation – that is the situation already dealt with. This means really a situation depending upon special values of the parameters. We confine the treatment to singularities at finite distance.

Let $\{\gamma_h\}$ be a complete family of independent quaternionic cycles. Upon separating the four quaternionic components for each cycle there will result a larger family $\{\gamma_{hj} | 1 \leq j \leq 4; 1 \leq h \leq \rho\}$ of rational cycles: with real rational coefficients. Let the rank of the matrix

$$\mu = [q_{hj}]$$

for all variable parameters be ρ. The singular locations will arise when the rank μ is below ρ. Let $\{M_s\}$ $1 \leq s \leq \sigma$ be the collection of the minors of μ of rank ρ. The singular situations correspond to the solutions of the algebraic system

$$M_s = 0, \quad 1 \leq s \leq \sigma.$$

However, this system contains other variables than the parameters. These extraneous variables may be eliminated by a standard algebraic method. There results a strict algebraic system

$$\mathscr{N}_u = 0, \quad 1 \leq u \leq \zeta,$$

consisting of polynomials in the parameters alone. The solutions of this system characterize a strict collection of singularities. Let V^μ denote its variety in the spaces of the parameters. The irreducible subvarieties $\{V^\nu\}$ of V^μ characterize the collection of all narrower singularities.

(7.1) Relation to the Landau variety. The Landau variety V^μ describes the projection of the real singular variety into the space Y of the parameters. Recall that it operates as follows: Take a point P in $Y - \text{proj.}\, V^\nu = Y_1$ and let Γ be the group of paths of Y_1. Let $\{\sigma_\alpha | 1 \leq \alpha \leq Q\}$ be the fundamental operations of Γ. To each σ_α there will correspond a residue $\hat{\sigma}_\alpha$ after the procedures described above. If the residue is zero set $\hat{\sigma}_\alpha = 1$. To an operation $\sigma_\alpha \sigma_{\alpha'} \cdots \sigma_\beta$ of Γ there corresponds the residue $\hat{\sigma}_\beta \cdots \hat{\sigma}_{\alpha'} \hat{\sigma}_\alpha$ of the integral (modulo a certain constant product). This gives an outline of the complete solution of the variation problem in the general case.

8. On the Loop-Complex

Let each loop λ of the complex G be viewed as an abstract branch. We have shown that any two loops λ, λ' have at most one node or one arc of G in common. If λ, λ' intersect one assigns to the pair an intersection point (λ, λ') of the pair of branches λ, λ'. There results a new complex $K(\lambda)$.

(8.1) *No collection of elements of* $K(\lambda)$ *makes up a loop of* $K(\lambda)$. *This implies that* $K(\lambda)$ *consists of a finite set of trees. Their number is the Betti number* $R_o(K)$.

Of course, this has no particular relation to the problem of Feynman integration.

9. Some Complements

They have been largely suggested in a letter from Professor Feynman. The first two are very simple, the third involves some slight complications.

(9.1) *It appears that many of the* m_h *may be simultaneously zero. This really merely means that* $\{\Sigma^m, \Sigma'^m\}$ *are just the two sheets of a cone of revolution whose base is an ordinary 2-sphere.*

(9.2) *It may also happen that some of the* m_h *are* $\neq 0$ *and equal: this does not affect the general argument.*

(9.3) *In the integral* (4.1) *there may well appear in the numerator of the integrand a factor, which in our present notation is a polynomial*

$$G(t_1, x_1, y_1, z_1, \ldots, t_R, \ldots, z_R).$$

This requires a slight additional argument. Let G_h denote what G becomes when all the variables except $t_h, \ldots, z_h$ are set equal to zero.

Now the varieties G^* and $Q = 0$ intersect in a surface Φ. Let P be a point of Φ. In the neighborhood of P the variety G^* is a power series in z the powers $t - t_o, \ldots, z - z_o$, where $t_o, \ldots, z_o$ are the coordinates of P. Hence $G^*(P) = 0$. It follows that all the residues as calculated earlier are multiplied by $G^*(P)$ and hence vanish at P. Therefore, along Φ there are no residues of the integral, and the latter is meromorphic about P.

10. Examples

(10.1) The representative figure of the first is Figure 10.

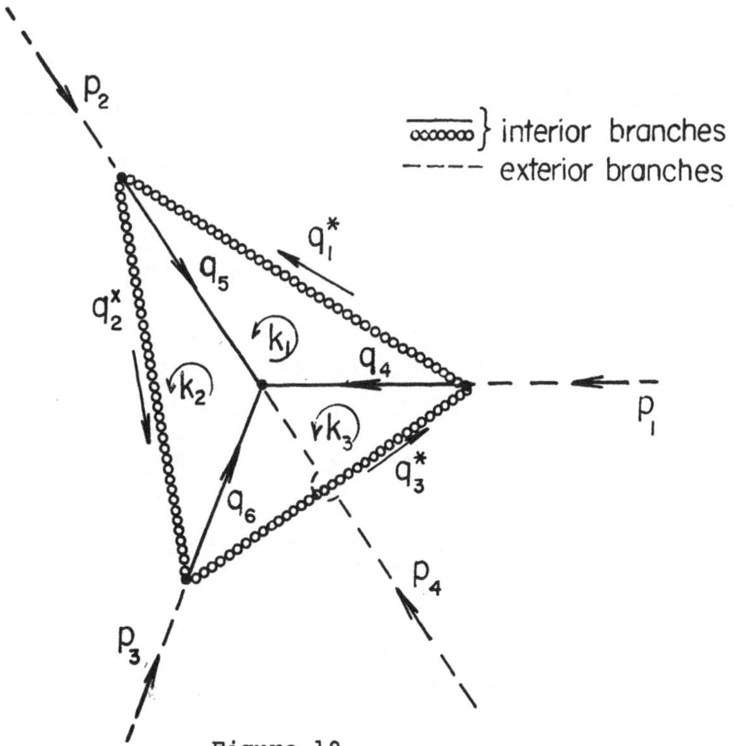

Figure 10.

Referring to this figure we arrive at the basic relations expressing all the q's in the terms of the k's and the p's:

$$q_h^* = k_h, \quad h = 1,2,3$$

$$q_4 = p_1 + k_3 - k_1$$

$$q_5 = p_2 + k_1 - k_2$$

$$q_6 = p_3 + k_2 - k_3$$

$$p_1 + p_2 + p_3 + p_4 = 0.$$

There are three loops $\lambda_1, \lambda_2, \lambda_3$. We examine the relations for λ_1 and deduce those for λ_2, λ_3 by obvious permutations of indices.

In conformity with our general designations

$$Q_1 = t_1^2 - x_1^2 - y_1^2 - z_1^2 - m_1^2.$$

For $m_1 \neq 0$ the locus $Q_1 = 0$ consists of a generalized hyperboloid of two sheets separated by and symmetrical with respect to $t_1 = 0$. The separation is complete when $m_1 \neq 0$ and the two sheets join when $m_1 = 0$ at the unique coordinate origin $t_1 = x_1 = \cdots = 0$. The coresponding variations are given earlier, and we need not repeat them.

The permutations are given by

$$1 \to 2 \to 3, \ 4 \to 5 \to 6.$$

This gives a complete description of the situation.

(10.2) The second example is described by Figure 11. It is shorter than the first but has no permutation symmetry. The basic

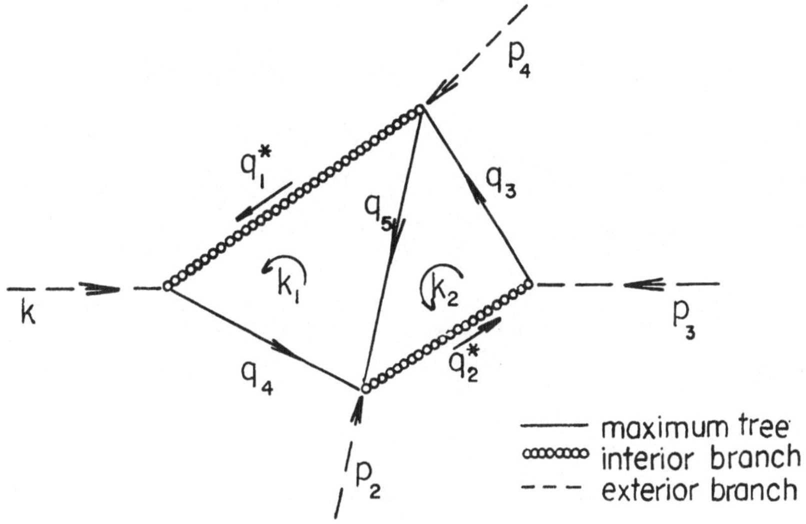

Figure 11.

relations are:

$$q_1^* = k_1$$

$$q_2^* = k_2$$

$$q_3 = k_2 + p_3$$

$$q_4 = k_1 + p_1$$

$$q_5 = k_1 + k_2 + p_1 - p_1.$$

There are two loops λ_1 and λ_2. We have

$$Q_i = k_i^* - m_i^2 = t_i^2 - x_i^2 - y_i^2 - z_i^2 - m_i^2, \quad i = 1,2.$$

The variations are easily read from earlier arguments.

11. Calculation of an Integral

We shall calculate the integral relative to the first example of Section 10 with a linear factor

$$G = at + bx + cy + dz$$

in the numerator. The constant coefficients and the masses m will be so chosen as to avoid a crossing of $G = 0$ and Σ^m, Σ'^m, which may manifestly be done. Note that since the Q's are quadratic in the coordinates the integral $\to 0$ at $\pm\infty$, thus avoiding a well known complication.

For t alone variable the partial integral corresponding to it is

$$I_t = \int \frac{dt(at+bx+cy+dz)}{t^2-x^2-y^2-z^2-m^2} .$$

The previous treatment of the "variation" applies also here. The integral has to be taken along the whole real line t and its residues calculated. Since x,y,z are arbitrary (but fixed) we have three integrals:

$$\int_{-\infty}^{t'-\varepsilon'} , \int_{t'+\varepsilon}^{t''-\varepsilon''} , \int_{t''+\varepsilon''}^{+\infty}$$

where $\varepsilon', \varepsilon''$ are arbitrarily small and ultimately $\to 0$. The quantities $t' \pm \varepsilon'$, $t'' \pm \varepsilon''$ are on the circumferences utilized in computing the residues. Between the first and second integral and the second and third there appear ultimately the residues relative to Σ^m and Σ'^m, both equal to $\frac{\pi i}{t}$. Hence the integral is in the limit

$$\int_{-\infty}^{t'-} + \int_{t'+}^{t''-} + \int_{t''+}^{+\infty} + \frac{2\pi i}{t} .$$

The same computation applies to the three integrals to x,y,z yielding say for x

$$\int_{-\infty}^{x'-} + \int_{x'+}^{x''-} + \int_{x''+}^{+\infty} - \frac{2\pi i}{x}$$

and the same for y,z with obvious modification.

For the total integral one must introduce x and the product of the residues. The required modifications are obvious.

12. A Final Observation

I return here to a relation between the complexes G and H to clear up a moot point left pending (Theorem 11.2).

First some simplifying notations. Denote the branches b_{ej} by β_j, $0 \leq j \leq N$. Call T_G the earlier maximal tree T of G and T_H one of H. Complete T_G with β_o and the exterior vertex n_o — this will be T_H. In fact

(12.1) <u>This</u> T_H <u>is actually a maximal tree of</u> H.

Since T_H may only be augmented by a b_h^* or by a β_j we must prove that one or the other addition T_H ceases to be a tree.

At all events T_H is manifestly a tree. If augmented by b_h^* it acquires the loop λ_h of $G \subset H$. Take now β_j. Let n_o', n_j' designate the nodes of β_o, β_j both in $G \subset H$. As nodes of T_G they may be joined by an arc θ_j of T_G. Hence $\beta_o \cup \theta_j \cup \beta_j = \mu_j$ is a loop of H. Hence T_H has the maximal tree property for T_H and (12.1) follows

Let $\vec{\mu}_j$ be the cycle of μ_j. $\vec{\lambda}_h, \vec{\mu}_j$ are the cycles of H which complement (in H) the branches b_h^* and β_j $(j > 0)$. Since no h_j contains a branch b_h^* we may state:

(12.2) <u>Theorem</u>. <u>No branch</u> b_h^* <u>contains a</u> p_j <u>or any other element of a</u> $\vec{\mu}_j$.

For the p_j's and $\vec{\mu}_j$ are entirely composed of elements of $T_G - G \cup b_{ej}$, while every b_h^* is exterior to T_G and has nothing in common with the p_j's.

The above theorem was implicit in the earlier treatment and is now proved explicitly.

CHAPTER V

FEYNMAN INTEGRALS. B.

1. Introduction

What follows is a continuation of the general theory developed in Chapter IV. This prolongation was more or less inspired by the excellent Paris Thèse of Felix Pham. Briefly speaking it deals with the subgraphs, open or closed, of the basic graph G. The geometric methods of Chapter IV continue to prevail. For I continue to believe that they are more suitable for an Introduction to our theory, more so indeed than the very abstruse methods utilized by all previous authors.

2. General Theory

(2.1) In this and the next section I dwell upon ideas mostly developed between 1920 and 1930. For clarity it seems that it is best to start not with a mere graph but with a general simplicial complex.

(2.2) Let then $K = \{\sigma\}$ be a finite connected simplicial complex with the σ's as its simplexes. Except for the points they are all oriented and the dimension p is indicated by σ^p. The star $\mathrm{St}\,\sigma$ of σ is the collection of all σ^q (including σ^p) with σ^p as face. The boundary $B\sigma$ of σ is the set of all its proper faces and its closure $\mathrm{Cl}\,\sigma$ is $\sigma \cup B\sigma$.

All these elements are assumed geometric, K being in some Euclidean space.

If $\Sigma = \{\sigma^p\}$ is any subset of simplexes of K, I denote by $|\Sigma|$ the totality of points of the σ's $\in \Sigma$. In particular $|K|$ is a polyhedron with $|\mathrm{St}\,\sigma|$ for $\sigma \in K$ an open set of $|K|$ and $|\mathrm{Cl}\,\sigma|$ is a closed subset of the polyhedron $|K|$.

A closed subcomplex L of K is one such that $\sigma \in L \Longrightarrow \mathrm{Cl}\,\sigma \in L$. Thus $|L|$ is a closed subset of the polyhedron $|K|$. It follows that $L_1 = K - L$ is such that (a) $\sigma \in L_1 \Longrightarrow \mathrm{St}\,\sigma \in L_1$ hence (b) $|L_1|$ is an open subset of the polyhedron $|K|$.

Notice that the chain-cycle arrangements in L are the same as in K. A difference is presented by L_1.

3. Relative Theory

In L_1 we have to deal with a new facet of our theory. This arises as follows. The chains of L_1 are really incomplete in that their boundaries ∂c may have parts in L. This gives rise to the relative theory.

A chain c^p of L_1 with $\partial c^p \subset L$ is referred to as a p-cycle mod L or a relative p-cycle of L_1. We call $c^p = \gamma^p$ bounding cycle mod L whenever there is a c^{p+1} such that

$$\partial c^{p+1} = \gamma^p + d^p, \quad d^p \subset L.$$

The maximum number of linearly independent p-cycles of L_1 mod L is the Betti number of L_1 mod L. Actually we may neglect in the part in L the chain in $L^* = L \frown \mathrm{Cl}\,L$ and this is assumed henceforth. The relative cycles are easily seen to have all the usual cycle properties.

4. Application to Graphs

Let now $K = G$ be the same graph as before. Thus L is a closed subcomplex of G and L_1 is its open complement.

(4.1) In L then we are only concerned with chains of dimension zero and one and for the present only real chains. The only dimensions being 0 and 1, every c^1 is a one-cycle and none of these are bounding. Therefore the Betti number $R^1(L)$ is merely the maximum number of linearly independent one-cycles of L.

Regarding the zero-chains every c^o is a zero-cycle. The Betti number $R^o(L)$ is the maximum number of linearly independent zero-chains mod (bounding one-chains). The related Betti number $R^o(L)$ is here the number of components of L.

(4.2) <u>To conform with our treatment of</u> G <u>itself write</u> R^L <u>for</u> $R_1(L)$. <u>We have then all the elements for expressing the Feynman integrals relative to</u> L.

The quadratic polynomials Q_j^L are attached to the branches b^L of L. Since $L \subset G$ their number is smaller than the earlier number of Q's.

We may freely add to L the condition that it be connected. Referring to Chapter IV.A., (1.4), L will have a unique maximal tree T. Its complement in L consists of branches b_h^L, $1 \leq h \leq R^L$. They have currents q_h^{*L} flowing in loops μ_h, and we have all that is required to write down the Feynman integral attached to L.

(4.3) It has been assumed throughout that G and all its subcomplexes are closed. However, the literature repeatedly envisages possible open subcomplexes. They do, in fact, open new perspectives. Consider then the open complement L_1 of L in G. What is required is to define an analogue T^* of a maximal tree. It would be a connected graph deprived of an absolute or relative one-cycle. Now a relative one-cycle γ^1 is a chain with boundary a zero-chain c^o in $L \cap \mathrm{Cl}\, L_1 = L^*$. This last complex would have a homology group H^o with a finite number of base elements $\{c_h^o \mid 1 \leq h \leq \zeta\}$. Thus the relative cycles have for base a collection $\{\gamma_h^1\}$ where

$$\partial \gamma_h^1 = c_h^o .$$

Necessary conditions are that T^* be a connected open subgraph of L_1, and contain no linear combination of the γ_h^1, in particular it must contain no absolute one-cycle. A maximal T^* is one which ceases to be a T^* upon being augmented by a branch of L_1. The complement of T^* in L_1 is again a finite collection of disjoint branches $\hat{b}_h$ of L_1.

5. On Certain Transformations

(5.1) Mapping $G \to L$. From the inclusion i: $L \xrightarrow{i} G = L \cup L_1$ we may define

$$G = L \cup L_1 \to L$$

identity $L \to L$; $L_1 \to 0$. This is in a sense an analytic mapping $G \to L$ such that if $\underline{t}$ is its "vector-variable" then the mapping is analytic with L as a singular locus. This may be compared with an analytic mapping $G \to L$ which has L as singular position, and considered by Speer [12, II]. Needless to say Speer's mapping is far more sophisticated than the one just presented.

(5.2) Mapping $L_1 \to G$. The result just given will help to orient the reader and details may well be omittted.

BIBLIOGRAPHY

Hwa, R. C. and V. L. Teplitz

[1] Homology and Feynman Integrals, Benjamin, New York, 1966.

Lefschetz, Solomon

[2] Topology, Am. Math. Soc. Colloquium Publ. 12, 1930. Reprinted by Chelsea, 2nd ed., 1953.

[3] L'Analysis Situs et la Géometrie Algébrique, Gauthier-Villars, Paris, 1924.

Leray, Jean

[4] Le calcul différential et intégral sur une variété analytique complexes, Bull. Soc. Mathématique de France, 87, 1959.

Pham, Frédéric

[5] Introduction à l'étude Topologique des Singularitiés de Landau, Gauthier-Villars, Paris, 1967.

Picard, Émile

[6] Traite d'analyse, Vol. 2, Gauthier-Villars, Paris.

Picard, Emile and George Simart

[7] Théorie des Fonctions Algébriques de Deux Variables, Vol. 2, Gauthier-Villars, Paris.

Poincaré, Henri

[8] Sur les résidues des intégrales doubles, Acta Math. 9, 320-380, Stockholm, 1887.

Regge, Tullio

[9] Algebraic topology methods in the theory of Feynman relativistic amplitudes, Battelle Memorial Institute, Recontres, 1967.

van der Waerden, B. L.

[10] Einfübrung Über Algebraische Geometrie, Teubner, Leipzig, 1939.

Streater, R. F. and A. S. Wightman

[11] Spin and Statistics and All That, Benjamin, New York, 1964

Speer, E. R.

[12] Generalized Feynman amplitudes, Annals of Math. Study No. 62, Princeton University Press, Princeton, N. J.

PART I

SUBJECT INDEX

affine space 15, 16

arc 28
  of graph 35-36

Ayres 100

base
  of one-cycles 48, 55
  of vector space 8

Betti numbers
  of complex 68-69
  of graph 40-41
  of Mayer sequence 19
  subdivision invariance 69
  of surface 76

Boundary relation
  for chains 44
  for complex 67

branch 34
  boundary relation of 44
  incidence number 44
  orientation of 43

Brayton 4

capacitor 51

cell 28

chain
  boundary relation of 44
  orientable 74-75

characteristic
  of complex 62, 66
  of graph 39, 41
  of Mayer sequence 19
  of projective plane 84
  of sphere 83
  subdivision invariance of 66
  of surface 83

closed set 28

closure 32

coboundary 49
  as voltage distribution 53

cochain 49

cocycle 49

Cohomology 50

compactness 32

complement 28

complex
  Betti numbers of 68-69
  boundary operator of 67
  characteristic of 62, 66
  connectedness 63
  Homology of 68-69
  incidence matrix of 68
  incidence numbers of 68
  orientation of 67
  polyhedron of 63
  subdivision of 64-66, 68

component 29
  of graph 36

connectedness 29
  of complex 63

continuity 27

current distribution 51-52, 103-1
  as cycle 52

cut 78

cycle 44
  base for 48, 55
  as current distribution 52
  of forest 47
  of loop 45, 47
  on surface 75

differential equation of electrical network 56-59

dimension
  of Cohomology 50
  of Homology 46-48
  of vector space 8

direct sum 9

double covering 76

duality 13, 15-17
  dual transformation 17
  of electrical network 104-106
  for graphs 48-50
  of Mayer sequence 20
  in surfaces 86-87
  symmetry of 18

dual space 15

dual transformation 16

electrical network 51
  differential equation of 56-59
  duality in 104-106
  reciprocal 103-104

electrostatic potential 53

Euclidean space 26

factor space 10

field 7, 67, 69

forest 36
  cycles in 47
  in electrical networks 54-55

genus 83

graph
  arc of 35-36
  Betti numbers of 40-41
  branch of 34
  chains of 43-50
  characteristic of 39, 41
  component of 36
  cotheory for 48-50
  cycles of 44-50
  forest 36
  incidence matrix 44
  loop in 34
  maximal tree of 39, 41
  node of 34, 36-37
  order of a point of 34
  orientation of 43
  Planar 89
  polyhedron of 36
  separable 91
  spherical 90
  subdivision of 38
  topological invariants 38-42
  tree 36-37

homeomorphism 24, 27

Homology
  of complex 68-69
  dimension of 46-48
  subdivision invariance of 68-69

incidence matrix
  of complex 68
  of graph 44

incidence number
  of complex 68
  of graph 44

inductor 54

intersection 27

interval 29

into 27

invariants
  subdivision 66, 68-69
  topological 30, 38-42 73, 83-84

inverse
  of matrix 6
  of transformation 27

isomorphism 8

Jordan 30
  curve 30

Jordan-Schoenflies Theorem 30, 79

Kirchoff 51

Kirchoff's Laws 51-52

Kronecker Delta 14

Kronecker Index 16

Kuratowski 89, 96

limit point 32

linear function 15

linear independence 8

linear transformation 16
  dual of 17
  nucleus of 17

loop 35
  cycles of 45, 47
  orientation of 45

MacLane 89, 91

mapping 27

matrix 5-7
  incidence matrix 44

Mayer sequence 18
  Betti numbers of 19
  characteristic 19
  dual of 20

Möbius strip 25
  in projective plane 84

Moser 3

neighborhood 28

node 34, 36-37
  boundary relation 44
  incidence number 44

nucleus 17

onto 26

open set 28

order of a point 34

orientation 25
  and Betti numbers 76
  of branch 43
  of complex 67
  of graph 43
  invariance of 73
  of loop 45
  of surface 72-78

Poincaré 13

polyhedron
  of complex 63
  of graph 36

potential, electrostatic 53

projective plane 13
  characteristic 84
  covering surface 85
  orientability 84

rank 6

resistor 53-54

Schoenflies 31

segment 29

set 27-28

simplex 61

sphere 28
  characteristic of 83
  orientation 79

spheroid 28

subdivision
  barycentric 66
  elementary 38, 64-65, 69
  invariant 66, 68-69
  of surface 79

surface
  Betti number of 76
  characteristic of 83
  cycles on 75
  double covering 76
  duality in 86-87
  genus of 83
  normal forms of 83-86
  orientation of 72-75
  subdivision of 79

Sylvester 12

topological invariants 30
  of graph 38-42
  of surface 73, 83-84

Topology 24, 27

transformation 14
  dual 17
  into 27
  linear 16
  onto 26
  topological 27

transpose 6

tree 36-37
  maximal 39, 41

Triangle Law 26

umbrella 71

union 27

van der Pol 54

Veblen 31

vectors, vector spaces
  base for 8
  column vector 10
  dimension 8
  direct sum 9

dual space 15
  factor space 10
  isomorphism 8
  linear function 15
  linear independence 7
  linear transformation 16

voltage distribution 52-53, 103-104
  as coboundary 53

Whitney 89

PART II

SUBJECT INDEX

Abelian integral 113, 135

Alexander 128

algebraic
  hypersurface 121
  surface 113, 152
  variety 115, 121-123, 149

arc 154, 158

Betti numbers 125, 133
  of graph 157
  of Φ_z 144
  subdivision 128

boundary 125
  in graph 157-159
  relative 126, 177-180
  singular 129

branch
  curve 141
  point 137, 138

Carrier 129

chain 125
  boundary of 125
  in graph 156-160
  intersection 131-134
  singular 129

characteristic
  of complex 125-126
  of graph 157

complex 115, 124
  characteristic of 125-126
  covering complex 117, 121, 150
  Poincaré group of 131
  subcomplex 126, 127
  subdivision of 127, 128

current 158, 161, 163

cycle 125
  in graph 157, 160
  intersection of 131-134
  in lacet graph 144
  in M 147-148
  of Φ_z 144
  quaternion 161, 168
  relative 126, 177-180
  singular 129

dimension of variety 122

duality
  Lefschetz 127
  in M 147
  Poincaré 127

Feynman 114

Feynman integral 154, 162, 174, 1…
  equal masses in 167
  graph of 160, 175-176
  variation in 163-168

forest, see tree, maximal

form 121

Fuchs 136

function field 122

generic point 122

graph 154
  characteristic of 157
  Feynman 160
  inseparable 154
  orientation of 156

group, Poincaré 130-131, 151, 166

Homology 125, 133
  relative 126, 177-180
  singular 129-130
  and subdivision 128

Homotopy 128

hypersurface, algebraic 121

incidence
  matrix 157
  number 157

intersection
  of chains 131-135
  in M 147
  number 132, 144, 147
  in Φ_z 144

junction 154

Kirchoff's Law 157, 161

lacet
  construction 138-139, 143, 152
  cycles on 144
  graph of 143

Landau variety 148, 169

Lefschetz, Picard-L. Theorem 113, 140, 148

loop 154, 158-161

M (manifold generated by Φ_z) 146-148

manifold 115
  absolute 126
  duality in 127
  generated by Φ_z 146-148
  orientation of 119, 127
  projective space 119-121

masses, in Feynman integral 162
  equal 167, 170
  zero 170

Menger-Urysohn dimension 122

mesh 126

orientation
  of graph 156
  of manifold 119
  of simplex 125

Pham 177

Φ_z, see Riemann surface

Picard 113, 114, 135-136, 152

Picard-Lefschetz Theorem 113, 140, 148

pinch point 138

Poincaré 127, 130, 150
  group 130-131, 151, 166

polar locus 164-166

Puiseux 141

quaternion 161, 162
  cycle 161, 168

relative homology 126, 178-181
  singular 129

residue 165-166, 169

Riemann surface (Φ_z) 138-139, 143
  Betti numbers of 144
  branch curves in 141
  cycles of 144
  intersection numbers in 144
  lacets in 138-139, 143
  manifold generated by 146-147

simplex 124
  orientation of 125
  singular 129-130
  star of 126

simplicial complex, see complex

simplicial map 128

Singular Homology 129-130

singularity 113-114, 123
159, 163

polar locus 169-165

Speer 180

star 126, 177

subdivision 127-128

surface, algebraic 113, 152

tree 154-155

maximal 155-156, 175-176
179-180

unit 115

irreducible non-unit 116

Urysohn, Menger-U. dimension 122

variation

of cycles 140, 144-146

in Feynman integral 163-168

in higher dimensions 152-153

of integrals 140, 144-146

of lacets 145-146

in M 150

variety

algebraic 115, 121-122, 149

covering complex of 117

dimension of 122

function field of 122

generic points of 122

irreducible 122

Landau 149, 169

singularities of 123

Weierstrass 116

preparation theorem 116, 136-137

Applied Mathematical Sciences

Vol. 1 F. John
Partial Differential Equations
Second edition
ISBN 0-387-90111-6

Vol. 2 L. Sirovich
Techniques of Asymptotic Analysis
ISBN 0-387-90022-5

Vol. 3 J. Hale
Functional Differential Equations
ISBN 0-387-90023-3

Vol. 4 J. K. Percus
Combinational Methods
ISBN 0-387-90027-6

Vol. 5 R. von Mises and K. O. Friedrichs
Fluid Dynamics
ISBN 0-387-90028-4

Vol. 6 W. Freiberger and U. Grenander
A Short Course in Computational Probability and Statistics
ISBN 0-387-90029-2

Vol. 7 A. C. Pipkin
Lectures on Viscoelasticity Theory
ISBN 0-387-90030-6

Vol. 8 G. E. O. Giacaglia
Perturbation Methods in Non-Linear Systems
ISBN 0-387-90054-3

Vol. 9 K. O. Friedrichs
Spectral Theory of Operators in Hilbert Space
ISBN 0-387-90076-4

Vol. 10 A. H. Stroud
Numerical Quadrature and Solution of Ordinary Differential Equations
ISBN 0-387-90100-0

Vol. 11 W. A. Wolovich
Linear Multivariable Systems
ISBN 0-387-90101-9

Vol. 12 L. D. Berkovitz
Optimal Control Theory
ISBN 0-387-90106-X

Vol. 13 G. W. Bluman and J. D. Cole
Similarity Methods for Differential Equations
ISBN 0-387-90107-8

Vol. 14 T. Yoshizawa
Stability Theory and the Existence of Periodic Solutions and Almost Periodic Solutions
ISBN 0-387-90112-4

Vol. 15 M. Braun
Differential Equations and Their Applications
ISBN 0-387-90114-0

Vol. 16 S. Lefschetz
Applications of Algebraic Topology
ISBN 0-387-90137-X